Resch · Bönisch · Benkert · Furui · Seo · Bez (Eds.)
High Performance Computing on Vector Systems

Michael Resch · Thomas Bönisch · Katharina Benkert
Toshiyuki Furui · Yoshiki Seo · Wolfgang Bez
Editors

High Performance Computing on Vector Systems

Proceedings of the High Performance Computing Center
Stuttgart, March 2005

With 128 Figures, 81 in Color, and 31 Tables

Springer

Editors

Michael Resch
Thomas Bönisch
Katharina Benkert

Höchstleistungsrechenzentrum
Stuttgart (HLRS)
Universität Stuttgart
Nobelstraße 19
70569 Stuttgart, Germany

resch@hlrs.de
boenisch@hlrs.de
benkert@hlrs.de

Wolfgang Bez

NEC High Performance
Europe GmbH
Prinzenallee 11
40459 Düsseldorf, Germany

wbez@hpce.nec.com

Toshiyuki Furui

NEC Corporation
Nisshin-cho 1-10
183-8501 Tokyo, Japan

t-furui@bq.jp.nec.com

Yoshiki Seo

NEC Corporation
Shimonumabe 1753
211-8666 Kanagawa, Japan

y-seo@ce.jp.nec.com

Front cover figure: Image of two dimensional magnetohydrodynamics simulation where current density has decayed from an Orszag-Tang vortex to form cross-like structures

Mathematics Subject Classification (2000): 65-06, 68U20, 65C20

ISBN-13 978-3-662-50045-3 ISBN 3-540-35074-8 (ebook)
DOI 10.1007/3-540-35074-8 Springer Berlin Heidelberg New York

Springer is a part of Springer Science+Business Media

springer.com

Softcover reprint of the hardcover 1st edition 2006

Typeset by the editors using a Springer TEX macro package
Production and data conversion: LE-TEX Jelonek, Schmidt & Vöckler GbR, Leipzig
Cover design: *design & production* GmbH, Heidelberg

Printed on acid-free paper 46/3142/YL - 5 4 3 2 1 0

Preface

In March 2005 about 40 scientists from Europe, Japan and the US came together the second time to discuss ways to achieve sustained performance on supercomputers in the range of Teraflops. The workshop held at the High Performance Computing Center Stuttgart (HLRS) was the second of this kind. The first one had been held in May 2004. At both workshops hardware and software issues were presented and applications were discussed that have the potential to scale and achieve a very high level of sustained performance.

The workshops are part of a collaboration formed to bring to life a concept that was developed in 2000 at HLRS and called the "Teraflop Workbench". The purpose of the collaboration into which HLRS and NEC entered in 2004 was to turn this concept into a real tool for scientists and engineers. Two main goals were set out by both partners:

- To show for a variety of applications from different fields that a sustained level of performance in the range of several Teraflops is possible.
- To show that different platforms (vector based systems, cluster systems) can be coupled to create a hybrid supercomputer system from which applications can harness an even higher level of sustained performance.

In 2004 both partners signed an agreement for the "Teraflop Workbench Project" that provides hardware and software resources worth about 6 MEuro (about 7 Million $ US) to users and in addition provides the funding for 6 scientists for 5 years. These scientists are working together with application developers and users to tune their applications. Furthermore, this working group looks into existing algorithms in order to identify bottlenecks with respect to modern architectures. Wherever necessary these algorithms are improved, optimized, or even new algorithms are developed.

The Teraflop Workbench Project is unique in three ways:

First, the project does not look at a specific architecture. The partners have accepted that there is not a single architecture that is able to provide an outstanding price/performance ratio. Therefore, the Teraflop Workbench is a hybrid architecture. It is mainly composed of three hardware components

- A large vector supercomputer system. The NEC SX-8/576M72 has 72 nodes and 576 vector processors. Each processor has a peak performance of 22 GFLOP/s which results in a peak overall performance of the system of 12.67 TFLOP/s. The sustained performance is about 9 TFLOP/s for Linpack and about 3–6 TFLOP/s for applications. Some of the results are shown in this book. The system is equipped with 9.2 TB of main memory and hence allows to run very large simulation cases.
- A large cluster of PCs. The 200 node system comes with 2 processors per node and a total peak performance of about 2.4 TFLOP/s. The system is perfectly suitable for a variety of applications in physics and chemistry.
- Two shared memory front end systems for offloading development work but also for providing large shared memory for pre-processing jobs. The two systems are equipped with 32 Itanium (Madison) processors and provide a peak performance of about 0.19 TFLOP/s each. They come with 0.256 TB and 0.512 TB of shared memory respectively which should be large enough even for larger pre-processing jobs. They are furthermore used for applications that rely on large shared memory such as some of the ISV codes used in automobile industry.

Second, the collaboration takes an unconventional approach towards data management. While mostly the focus is on management of data the Teraflop Workbench Project considers data to be the central issue in the whole simulation workflow. Hence, a file system is at the core of the whole workbench. All three hardware architectures connect directly to this file system. Ideally the user only once has to transfer basic input information from his desk to the workbench. After that data reside inside the central file system and are only modified either for pre-processing, simulation or visualization.

Third, the Teraflop Workbench Project does not look at a single application or a small number of well defined problems. Very often extreme fine-tuning is employed to achieve some level of performance for a single application. This is reasonable wherever a single application can be found that is of overwhelming importance for a centre. For a general purpose supercomputing centre like the HLRS this is not possible. The Teraflop Workbench Project therefore sets out to tackle as many fields and as many applications as possible. This is also reflected in the contents of this book. The reader will find a variety of application fields that range from astrophysics to industrial combustion processes and from molecular dynamics to turbulent flows. In total the project supports about 20 projects of which most are presented here.

In the following the book presents key contributions about architectures and software but many more papers were collected that describe how applications can benefit from the architecture of the Teraflop Workbench Project. Typically sustained performance levels are given although the algorithms and the concrete problems of every field still are at the core of each contribution.

As an opening paper NEC provides a scientifically very interesting technical contribution about the most recent system of the NEC SX family the SX-8. All of the projects described in this book either use the SX-8 system of HLRS as the simulation facility or provide comparisons of applications on the SX-8 and other systems. The paper can hence be seen as an introduction of the underlying hardware that is used by various projects.

In their paper about vector processors and micro processors Peter Lammers from the HLRS, Gerhard Wellein, Thomas Zeiser, and Georg Hager from the Computing Centre, and Michael Breuer from the chair for fluid mechanics at the University of Erlangen, Germany, look at two competing basic processor architectures from an application point of view. The authors compare the NEC SX-8 system with the SGI Altix architecture. The comparison is not only about the processor but involves the overall architecture. Results are presented for two applications that are developed at the department of fluid mechanics. One is a finite volume based direct numerical simulation code while the other is based on the Lattice Boltzmann method and is again used in direct numerical simulation. Both codes rely heavily on memory bandwidth and as expected the vector system provides superior performance. Two points are, however, very notable. First, the absolute performance for both codes is rather high with one of them reaching even 6 TFLOP/s. Second, the performance advantage of the vector based system has to be put into relation with the costs which gives an interesting result.

A similar but more extensive comparison of architectures can be found in the next contribution. Jonathan Carter and Leonid Oliker from Lawrence Berkeley National Laboratory, USA have done a lot of work in the field of architecture evaluation. In their paper they describe recent results on the evaluation of modern parallel vector architectures like the Cray X1, the Earth Simulator and the NEC SX-8 and compare them to state of the art microprocessors like the Intel Itanium the AMD Opteron and the IBM Power processor. For their simulation of magnetohydrodynamics they also use a Lattice Boltzmann based method. Again it is not surprising that vector systems outperform microprocessors in single processor performance. What is striking is the large difference which combined with cost arguments changes the picture dramatically.

Together these first three papers give an impression of what the situation in supercomputing currently is with respect to hardware architectures and with respect to the level of performance that can be expected. What follows are three contributions that discuss general issues in simulation – one is about sparse matrix treatment, a second is about first-principles simulation while the third tackles the problem of transition and turbulence in wall-bounded shear flow. All three problems are of extreme importance for simulation and require a huge level of performance.

Toshiyuki Imamura from the University of Electro-Communications in Tokyo, Susumu Yamada from the Japan Atomic Energy Research Institute (JAERI) in Tokyo, and Masahiko Machida from Core Research for Evolutional Science and Technology (CREST) in Saitama, Japan tackle the problem of condensation of fermions to investigate the possibility of special physical properties like superfluidity. They employ a trapped Hubbard model and end up with a large sparse matrix. By introducing a new preconditioned conjugate gradient method they are able to improve the performance over traditional Lanczos algorithms by a factor of 1.5. In turn they are able to achieve a sustained performance of 16.14 TFLOP/s on the earth simulator solving a 120-billion-dimensional matrix.

In a very interesting and well founded paper Yoshiyuki Miyamoto from the Fundamental and Environmental research Laboratories of NEC Corporation describes simulations of ultra-fast phenomena in carbon nanotubes. The author employs a new approach based on the time-dependent densitiy functional theory (TDDFT), where the real-time propagation of the Kohn-Sham wave functions of electrons are treated by integrating the time-evolution parameter. This technique is combined with a classical molecular dynamics simulation in order to make visible very fast phenomena in condensed matters.

With Philipp Schlatter, Steffen Stolz, and Leonhard Kleiser from the ETH Zürich, Switzerland we again change subject and focus even more on the application side. The authors give an overview of numerical simulation of transition and turbulence in wall-bounded shear flows. This is one of the most challenging problems for simulation requiring a level of performance that is currently beyond our reach. The authors describe the state of the art in the field and discuss Large Eddy Simulation (LES) and Subgrid-Scale models (SGS) and their usage for direct numerical simulation.

The following papers present projects tackled as part of the Teraflop Workbench Project.

Malte Neumann and Ekkehard Ramm from the Institute of Structural Mechanics in Stuttgart, Germany, Ulrich Küttler and Wolfgang A. Wall from the Chair for Computational Mechanics in Munich, Germany, and Sunil Reddy Tiyyagura from the HLRS present findings for the computational efficiency of parallel unstructured finite element simulations. The paper tackles some of the problems that come with unstructured meshes. An optimized method for the finite element integration is presented. It is interesting to see that the authors have employed methods to increase the performance of the code on vector systems and can show that also microprocessor architectures can benefit from these optimizations. This supports previous findings that cache optimized programming and vector processor optimized programming very often lead to similar results.

The role of supercomputing in industrial combustion modeling is described in an industrial paper by Natalia-Currle Linde, Uwe Küster, Michael Resch, and Benedetto Risio which is a collaboration of HLRS and RECOM Services – a small enterprise at Stuttgart, Germany. The quality of simulation in the optimum design and steering of high performance furnaces of power plants has reached a level at which it can compete with physical experiments. Such simulations require not only an extremely high level of performance but also the ability to do parameter studies. In order to relieve the user from the burden of submitting a set of jobs the authors have developed a framework that supports the user. The Science Experimental Grid Laboratory (SEGL) allows to define complex workflows which can be executed in a Grid environment like the Teraflop Workbench. It furthermore supports the dynamic generation of parameter sets which is crucial for optimization.

Helicopter simulations are presented by Thorsten Schwarz, Walid Khier, and Jochen Raddatz from the Institute of Aerodynamics and Flow Technology of the German Aerospace Center (DLR) at Braunschweig, Germany. The authors use a structured Reynolds-averaged Navier-Stokes solver to compute the flow field around a complete helicopter. Performance results are given both for the NEC SX-6 and the new NEC SX-8 architecture.

Hybrid simulations of aeroacoustics are described by Qinyin Zhang, Phong Bui, Wageeh A. El-Askary, Matthias Meinke, and Wolfgang Schröder from the Department of Aerodynamics of the RWTH Aachen, Germany. Aeroacoustics is a field that is getting important for aerospace industries. Modern engines of airplanes are so silent that the noise created from aeroacoustic turbulences has often become a more critical source of sound. The simulation of such phenomena is split into two parts. In a first part the acoustic source regions are resolved using a large eddy simulation method. In the second step the acoustic field is computed on a coarser grid. First results of the coupled approach are presented for relatively simple geometries. Simulations are carried out on 10 processors but will require much higher performance for more complex problems.

Albert Ruprecht from the Institute of Fluid Mechanics and Hydraulic Machinery of the University of Stuttgart, Germany, shows simulation of a water turbine. The optimization of these turbines is crucial to extract the potential of water power plants when producing electricity. The author uses a parallel Navier-Stokes solver and provides some interesting results.

A topic that is unusual for vector architectures is atomistic simulation. Franz Gähler from the Institute of Theoretical and Applied Sciences of the University of Stuttgart, Germany, and Katharina Benkert from the HLRS describe a comparison of an ab initio code and a classical molecular dynamics code for different hardware architectures. It turns out that the ab initio simulations perform excellently on vector machines. Again it is, however, worth to look at the ratio of performance on vector and microprocessor systems. The molecular dynamics code in its existing version is better suited for large clusters of microprocessor systems. In their contribution the authors describe how they want to improve the code to increase the performance also for vector based systems.

Martin Bernreuther from the Institute of Parallel and Distributed Systems and Jadran Vrabec from the Institute of Thermodynamics and Thermal Process Engineering of the University of Stuttgart, Germany, in their paper tackle the problem of molecular simulation of fluids with short range potentials. The authors develop a simulation framework for molecular dynamics simulations that specifically targets the field of thermodynamics and process engineering. The concept of the framework is described in detail together with algorithmic and parallelization aspects. Some first results for a smaller cluster are shown.

An unusual application for vector based systems is astrophysics. Konstantinos Kifonidis, Robert Buras, Andreas Marek, and Thomas Janka from the Max-Planck-Institute for Astrophysics at Garching, Germany, give an overview of the problems and the current status of supernova modeling. Furthermore they describe their own code development with a focus on the aspects of neutrino transports. First benchmark results are reported for an SGI Altix system as well as for the NEC SX-8. The performance results are interesting but so far only a small number of processors is used.

With the next paper we return to classical computational fluid dynamics. Kamen N. Beronov, Franz Durst, and Nagihan Özyilmaz from the Chair for Fluid Mechanics of the University of Erlangen, Germany, together with Peter Lammers from HLRS present a study on wall-bounded flows. The authors first present the state of the art in the field and compare different approaches. They then argue for a Lattice Boltzmann approach providing also first performance results.

A further and last example in the same field is described in the paper of Andreas Babucke, Jens Linn, Markus Kloker, and Ulrich Rist from the Institute of Aerodynamics and Gasdynamics of the University of Stuttgart, Germany. A new code for direct numerical simulations solving the complete compressible 3-D Navier-Stokes equations is presented. For the parallelization a hybrid approach is chosen reflecting the hybrid nature of clusters of shared memory machines like the NEC SX-8 but also multiprocessor node clusters. First performance measurements show a sustained performance of about 60% on 40 processors of the SX-8. Further improvements of scalability have to be expected.

The papers presented in this book provide on the one hand a state of the art in hardware architecture and performance benchmarking. They furthermore lay out the wide range of fields in which sustained performance can be achieved if appropriate algorithms and excellent programming skills are put together. As the first of books in this series to describe the Teraflop Workbench Project the collection provides a lot of papers presenting new approaches and strategies to achieve high sustained performance. In the next volume we will see many more results and further improvements.

Stuttgart, January 2006

M. Resch
W. Bez

Contents

Applications I: Finite Element Method

Applications II: Fluid Dynamics

Applications III: Particle Methods

Applications IV: Turbulence Simulation

List of Contributors

Future Architectures in Supercomputing

The NEC SX-8 Vector Supercomputer System

Satoru Tagaya[1], Masato Nishida[1], Takashi Hagiwara[1], Takashi Yanagawa[2], Yuji Yokoya[2], Hiroshi Takahara[3], Jörg Stadler[4], Martin Galle[4], and Wolfgang Bez[4]

[1] NEC Corporation, Computers Division,
1-10, Nisshin-cho, Fuchu, Tokyo, Japan,
[2] NEC Corporation, 1st Computers Software Division,
1-10, Nisshin-cho, Fuchu, Tokyo, Japan
[3] NEC Corporation, HPC Marketing Promotion Division,
1-10, Nisshin-cho, Fuchu, Tokyo, Japan,
[4] NEC High Performance Computing Europe GmbH,
Prinzenallee 11, D-40549 Düsseldorf, Germany

Abstract In 2003, the High Performance Computing Center in Stuttgart (HLRS) has decided to install 72 NEC SX-8 vector computer nodes with 576 CPUs in total. With this installation, the HLRS is able to provide the highest vector technology based computational power to academic and industrial users within Europe. In this article, an overview of the NEC SX-8 vector computer architecture is presented. After a general outline of the SX-8 series, a description of the SX-8 hardware is given. The article is finalized by an overview of related software features

1 Introduction

The SX-8 is the follow on system to the worlds most successful Vector Supercomputer system, the NEC SX-6 and SX-7 Series. The SX-8 system was announced in October 2004 and shipped to the first European customers in January of 2005. Like previous SX systems the SX-8 is designed for those applications which require the fastest CPU, the highest memory bandwidth, the highest sustained performance and the shortest time to solution available. Like its predecessors the SX-8 is completely air-cooled and based on state of the art CMOS-chip technology; beyond that, it incorporates novelties like highly sophisticated board and compact interconnect technologies.

At NEC, Tadashi Watanabe has led the design and strategy of the SX supercomputer line since the early 1980s. He has always focused on building vector supercomputers with extremely fast processors, the highest possible memory bandwidth and many levels of parallelism. By using less exotic and less costly technologies compared with other supercomputer designs, for example the introduction of complete air cooling starting with the SX-4, the manufacturing costs as well as costs for electricity have continuously been reduced with every

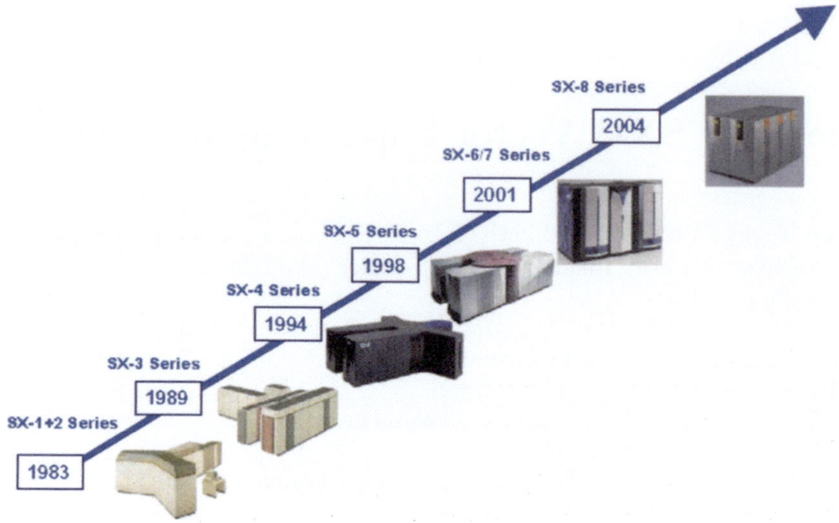

Fig. 1. NEC SX Product History

new generation of the SX series. Watanabe's basic design has produced one of the longest-lasting fully compatible HPC-product series ever built for the high performance computing market.

Watanabe has maintained the compatibility in the SX supercomputer line to protect customer investments in the SX product line. The investment cost of software is a major burden for most HPC users and a substantial cost for computer manufacturers, especially in porting, optimizing, and certifying third-party applications.

It is important to note that vector systems should not be viewed in opposition to parallel computing; vector computers implement parallelism at the fine-grained level through vector registers and pipelined functional units and at the medium-grained level through shared memory multiprocessor system configurations. In addition, these systems can be used as the basic building blocks for larger distributed memory parallel systems.

2 General Description of the SX-8 Series

NEC's latest approach to supercomputer architecture design is the combination of air-cooled CMOS processors with a multilayer PCB (printed circuit board) interconnect to build a wire-less single node. For the first time, the crossbar between CPUs and memory is implemented solely using a PCB. In all previous SX supercomputers, the interconnects were built using tens of thousands of cables between the processors, memory, and I/O. By moving to the PCB design, NEC was able to further increase the bandwidth with even lower latency while providing higher system reliability from the substantial decrease in hardware complexity. CMOS was chosen as the underlying basic technology because it offers substantial advantages over traditional ECL technologies in high performance circuit applications. Examples of these advantages include vastly reduced costs of manufacturing the basic VLSI (very large scale integrated) device due to fewer process steps, lower operational power consumption, lower heat dissipation and higher reliability because of the more stable technology and reduced parts counts enabled by the very large scale circuit integration.

By keeping the instruction set and software compatibility with the previous versions of the SX product line, customers can move their applications to the SX-8 system without having to rewrite or recompile those applications. This provides the SX-8 with the complete application set that has been developed and optimized over the past 20 years for the SX product line.

SX-8 Series systems are equally effective in general purpose or dedicated applications environments and are particularly well suited for design and simulation in such fields as aerospace, automotive, transportation, product engineering, energy, petroleum, weather and climate, molecular science, bio-informatics, construction and civil engineering.

SX-8 Product Highlights

- 16 or 17.6 GFLOPS peak vector performance, with eight operations per clock running at 2.0 or 2.2 GHz (0.5 or 0.45 ns cycle time); 1 or 1.1 GHz for instruction decoding/issuing and scalar operations
- Up to 8 CPUs per node, each single chip CPU manufactured in 90 nm Cu technology
- Up to 16 GB of memory per CPU, 128 GB in a single 8-way SMP node
- Up to 512 or 563.2 GB/s of memory bandwidth per node, 64 or 70.4 GB/s per CPU
- IXS Super-Switch between nodes, up to 512 nodes supported
- 16 or 32 GB/s bidirectional inter-node bandwidth (8 or 16 GB/s for each direction)
- running the mature SUPER-UX, System V port, 4.3 BSD with new enhancements for Multi Node systems; ease of use; support for new languages and standards; and operational improvements

The SX-8 Series continue to provide users with a high performance product which supports a physically shared and uniform memory within a node. The proven SX shared memory parallel vector processing architecture, a highly developed and reliable architecture enables users to efficiently solve their engineering and scientific problems. As with previous generation SX Series systems, these new generations provide ease of programming and allow for advanced automated vectorization and parallelization by the compilers.

SX-Series systems provide an excellent commercial quality, fully functional, balanced system capable of providing solutions for a broad range of applications requiring intensive computation, very large main memories, very high bandwidth between main memory and CPUs and very high input-output rates.

Table 1. SX-8 Series Model Overview

	FCRAM		DDR2-SDRAM	
	SX-8/A	SX-8/M	SX-8/A	SX-8/M
CPUs	4 - 8	8 - 4096	4 – 8	8 - 4096
Peak/CPU	16 GFLOPS	16 GFLOPS	17.6 GFLOPS	17.6 GFLOPS
Peak Performance	64 - 128 GFLOPS	128 - 65536 GFLOPS	70.4 – 140.8 GFLOPS	140.8 - 72090 GFLOPS
Clock	2 GHz	2 GHz	2.2 GHz	2.2 GHz
MMU				
Capacity	32 - 64 GB	64 GB - 32 TB	32 – 128 GB	128 GB - 64 TB
Bandwidth	256 - 512 GB/s	512 - 262144 GB/s	281.6 – 563.2 GB/s	563.2 - 288358 GB/s
IOF				
Max. number of I/O channels	55	28160	55	28160
Maximum data transfer rate	12,8 GB/s	6553 GB/s	12,8 GB/s	6553 GB/s

SX-8 Series models are designed to provide industry leading sustainable performance on real world applications, provide extremely high bandwidth computing capability and provide leading I/O capability in both capacity and single stream bandwidth. SX-8 Single Node models will typically provide 35–80% efficiency, enabling sustainable performance levels three to five times that of a highly parallel system build with workstation technology.

The SX-8 Series provides FORTRAN and C as well as C++ compilers with a high level of automatic vectorization and parallelization. Distributed memory parallel systems require use of programmer coded message passing and associated data decompositions or the use of HPF parallelization paradigm.

The SX-8 Series advantage is a true high end supercomputer system that outperforms HPC systems which are based on workstation technology, in terms of cost of system, cost of operation and total system reliability while providing leading supercomputer performance. Further, in cases where programming considerations must be accounted for, an SX-8 Series system could easily result in the lowest total cost solution because of the considerably reduced application development time enabled by shared memory programming models and automated vectorization and parallelization.

3 Technology and Hardware Description

The SX-8 Series was designed to take advantage of the latest technology available. The architecture combines the best of the traditional shared memory parallel vector design in Single Node systems with the scalability of distributed memory architecture in Multi Node systems. The usefulness of the architecture is evident as most modern competing vendors have adopted a similar architectural approach.

The SX-8 Series inherits the vector-processor based distributed shared memory architecture which was highly praised in the SX-5/SX-6 Series and flexibly works with all kinds of parallel processing schemes. Each shared memory type single-node system contains up to 8 CPUs (16 or 17.6 GFLOPS peak per CPU) which share a large main memory of up to 128 GB. Two types of memory technology are available, FCRAM (Fast Cycle RAM) with up to 64 GB per node and DDR2-SDRAM with up to 128 GB per node. In a Multi Node system configured with a maximum of 512 nodes, parallel processing by 4096 CPUs achieves peak performance of more than 65 TFLOPS. It provides a large-capacity memory of up to 65 TB with DDR2-SDRAM.

Through the inheritance of the SX architecture, the operating system SUPER-UX maintains perfect compatibility with the SX-8 Series. It is a general strategy of NEC to preserve the customers investment in application software.

3.1 SX-8 Series Architecture

All SX-8 models can be equipped with fast cycle or large capacity memory. Fast cycle memory (FCRAM) has faster access and lower latency time compared to DDR2-SDRAM whereas the memory bandwidth per FCRAM node is 512 GB/s and the memory bandwidth per DDR2-SDRAM node is 563 GB/s. On a per CPU basis 8 GB fast cycle memory can be configured whereas 16 GB large capacity memory can be configured. This provides a memory capacity up to 32 TB of fast cycle memory or 64 TB of large capacity memory for the 4096 processor SX-8 system. Figure 2 shows a multi node system. Details of the single node SX-8 system are illustrated in Fig. 3.

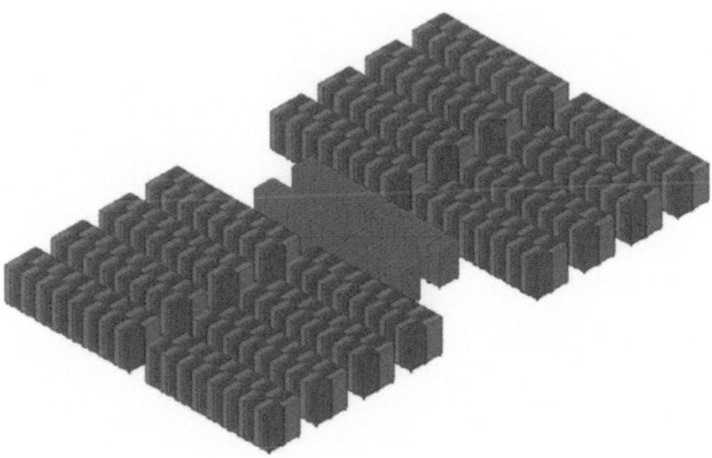

Fig. 2. SX-8/2048M256 system

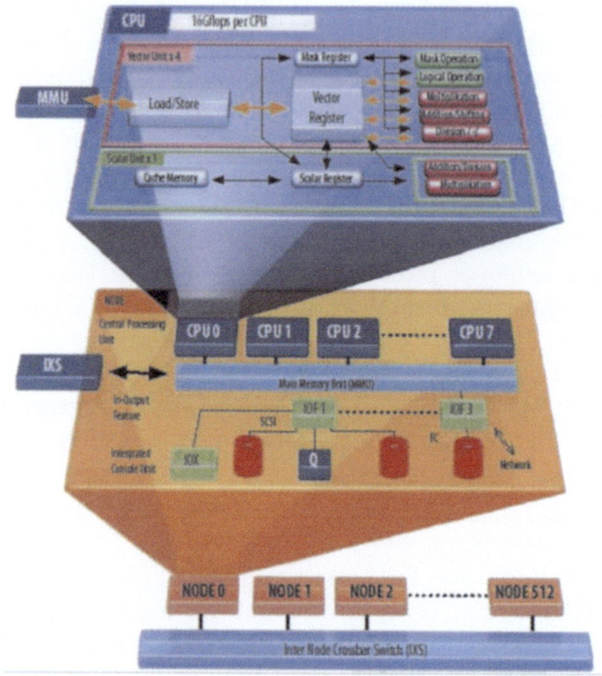

Fig. 3. Exploded view of the SX-8 system

3.2 SX-8 Single Node Models

The crossbar between CPUs and memory is implemented using a PCB for the first time. In all previous SX supercomputers, the interconnects were built using cables between the processors, memory, and I/O. By moving to a PCB design, about 20000 cables could be removed within one node providing higher reliability.

3.3 SX-8/M Series Multi Node Models

Multi Node models of the SX-8 providing up to 70.4 TFLOPS of peak performance (64 TFLOPS for the FCRAM version) on 4096 processors are constructed using the NEC proprietary high speed single-stage crossbar (IXS) linking multiple Single Node chassis together (architecture shown in Fig. 4). The high speed inter-node connection provides 8 or 16 GB/s bidirectional transfers and the IXS crossbar supports an 8 TB/s bisection bandwidth for the maximum of 512 nodes. Table 3 includes specifications of representative SX-8/M Multi Node FCRAM models. Multi Node models of the DDR2-SDRAM can be configured similarly.

3.4 Central Processor Unit

The central processing unit (CPU) is a single chip implementation of the advanced SX architecture which especially on vector codes achieves unrivaled efficiencies. In addition the CPU is equipped with a lot of registers for scalar

Table 2. SX-8 Single Node Specifications

Model group	SX-8/A FCRAM			SX-8/A DDR2-SDRAM		
Model name	8A	6A	4A	8A	6A	4A
Central processing unit (CPU)						
Number of CPUs	8	6	4	8	6	4
Peak Performance	128 GFLOPS	96 GFLOPS	64 GFLOPS	140.8 GFLOPS	105.6 GFLOPS	70.4 GFLOPS
Vector registers	144 kB x 8	144 kB x 6	144 kB x 4	144 kB x 8	144 kB x 6	144 kB x 4
Scalar registers	64bits x 128 regs x 8	64bits x 128 regs x 6	64bits x 128 regs x 4	64bits x 128 regs x 8	64bits x 128 regs x 6	64bits x 128 regs x 4
Main memory unit (MMU)						
Capacity	32, 48, 64 GB			32, 64, 96, 128 GB		
Max.data transfer Performance	512 GB/s	384GB/s	256 GB/s	563.2 GB/s	422.4 GB/s	281.6 GB/s
I/O devices						
Host channel adapters	2 – 4			2 - 4		
Max. number of PCI-X slots	55			55		
Max. data Transfer rate	12,8 GB/s read or write			12,8 GB/s read or write		

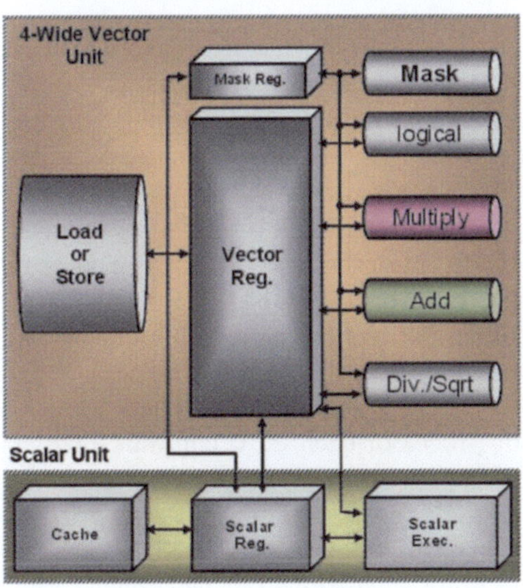

Fig. 4. CPU architecture of SX-8

arithmetic operations and base-address calculations so that scalar arithmetic operations can be performed effectively.

Each vector processor reaches a peak performance of 16 or 17.6 GFLOPS with the traditional notion of taking only add and multiply operations into account, neglecting the fully independent hardware divide/square-root pipe and the scalar units. The technological breakthrough of the SX-6 and SX-8 compared to previous generations is that the CPU has been implemented on a single

Table 3. SX-8/M DDR2-SDRAM Representative Systems Specifications

Model group	SX-8/M DDR2-SDRAM				
Model name	4096M512	256M32	128M16	64M8	16M2
Central processing unit (CPU)					
Number of CPUs	4096	256	128	64	16
Peak Performance	70.4 TFLOPS	4.4 TFLOPS	2.2 TFLOPS	1.1 TFLOPS	281.6 GFLOPS
Main Memory unit (MMU)					
Max. Capacity	32 TB	2 TB	1 TB	512 GB	128 GB
Max. Memory Bandwidth	281.6 TB/s	17.6 TB/s	8.8 TB/s	4.4 TB/s	1.1 TB/s

Table 4. SX-8/M FCRAM Representative Systems Specifications

Model group	SX-8/M FCRAM				
Model name	4096M512	256M32	128M16	64M8	16M2
Central processing unit (CPU)					
Number of CPUs	4096	256	128	64	16
Peak Performance	64 TFLOPS	4 TFLOPS	2 TFLOPS	1024 GFLOPS	256 GFLOPS
Main Memory unit (MMU)					
Max. Capacity	32 TB	2 TB	1 TB	512 GB	128 GB
Max. Memory Bandwidth	256 TB/s	16 TB/s	8 TB/s	4 TB/s	1 TB/s

chip. The 16 or 17.6 GFLOPS peak performance is achieved through add and multiply pipes which consist of 4 vector pipelines working in parallel on one single instruction. Taking into account the floating point vector pipelines with add/shift and multiply, this means that every clock cycle 8 results are produced. The major clock cycle of the SX-8 is 0.5 or 0.45 nsec, thus the vector floating point peak performance of each processor is 16 or 17.6 GFLOPS, respectively. The processors consists of vector add/shift, vector multiply, vector logical and vector divide pipelines. The vector divide pipeline which also supports vector square root, generates 2 results every second clock cycle leading to additional 4 or 4.4 GFLOPs. In addition to the vector processor each CPU contains a super-scalar unit.

The scalar unit runs on 1.0 or 0.9 ns clock speed. This processor is a 4-way super-scalar unit controlling the operation of the vector processor and executing scalar instructions. It executes 2 floating point operations per clock cycle, thus runs at 2 or 2.2 GFLOPS. Adding up the traditional peak performance of 16 or 17.6 GFLOPS, the divide peak performance of 4 or 4.4 GFLOPS and the scalar performance of 2 or 2.2 GFLOPS, each processor can achieve a maximum CPU performance of 22 or 24.2 GFLOPS, respectively.

The vector processor contains 16 KB of vector arithmetic registers which feed the vector pipes as well as 128 KB of vector data registers which are used to store intermediate results and thus avoid memory bottlenecks. The maximum bandwidth between each SX-8 CPU and the shared memory is 64 GB/s for the FCRAM version and 70.4 GB/s for the DDR2-SDRAM version.

One of the highlights of the SX-8 CPU architecture is the new MMU Interface Technology. The CPUs within one node are connected with low-latency so-called SerDes (Serializer-Deserializer) integration saving internal 20000 cables compared to the predecessor SX-6.

3.5 Parallel Pipeline Processing

Substantial effort has been made to provide significant vector performance for short vector lengths. The crossover between scalar and vector performance is a short 14 elements in most cases.

The vector unit is constructed using NEC vector pipeline processor VLSI technology. The vector pipeline sets comprise 16 individual vector pipelines arranged as sets of 4 add/shift, 4 multiply, 4 divide, and 4 logical pipes. Each set of 4 pipes services a single vector instruction and all sets of pipes can operate concurrently. With a vector add and vector multiply operating concurrently the pipes provide 16 or 17.6 GFLOPS peak performance for the SX-8.

The vector unit has 8 vector registers for 256 words of 8 Bytes each from which all operations can be started. In addition there are 64 vector data registers of the same size which can receive results from pipelines concurrently and from the 8 vector registers; the vector data registers serve as a high performance programmable vector buffer that significantly reduces memory traffic in most cases.

3.6 Memory Bank Caching

One of the new features of the SX-8 Series is the Memory Bank cache (Fig. 5). Each vector CPU has 32 kB of memory bank cache exclusively supporting 8 bytes for each of the 4096 memory banks. It is a direct-mapped write-through cache decreasing the bank busy time from multiple vector access to the same address.

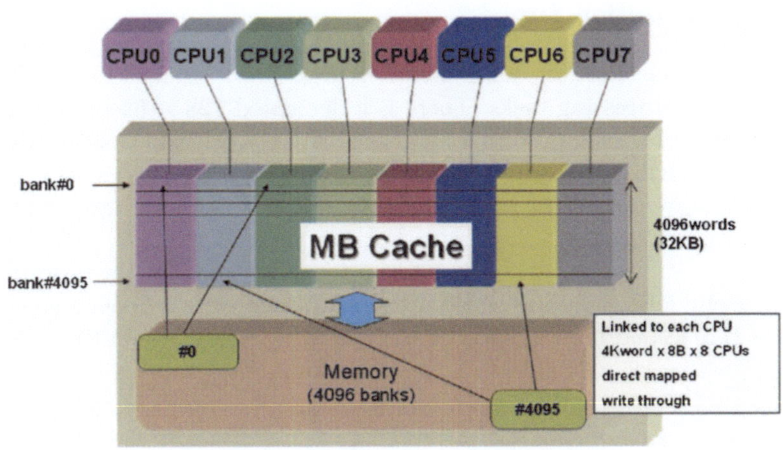

Fig. 5. Concept of Bank Caching in SX-8 node

On specific application this unique feature has proven to reduce performance bottlenecks caused by memory bank conflicts

3.7 Scalar Unit

The scalar unit is super-scalar with 64-kilobyte operand and 64-kilobyte instruction caches. The scalar unit has 128 × 64 bit general-purpose registers and operates at a 1 or 1.1 GHz clock speed for the SX-8. Advanced features such as branch prediction, data prefetching and out-of-order instruction execution are employed to maximize the throughput.

All instructions are issued by the super-scalar unit which can sustain decode of 4 instructions per clock cycle. Most scalar instructions issue in a single clock and vector instructions issue in two clocks. The scalar processor supports one load/store path and one load path between the scalar registers and scalar data cache. Furthermore, each of the scalar floating point pipelines supports both floating add, floating multiply and floating divide.

3.8 Floating Point Formats

The vector and scalar units support IEEE 32 bit and 64 bit data. The scalar unit also supports extended precision 128 bit data. Runtime I/O libraries enable reading and writing of files containing binary data in Cray[1] Research, IBM[2] formats as well as IEEE.

3.9 Fixed Point Formats

The vector and scalar units support 32 and 64 bit fixed point data and the scalar unit can operate on 8 and 16 bit signed and unsigned data.

3.10 Synchronization Support

Each processor has a set of communications registers optimized for synchronization of parallel processing tasks. There is a dedicated 128 × 64-bit communication registers set for each processor and each Single Node frame has an additional 128 × 64-bit privileged communication register set for the operating system.

Test-set, store-and, store-or, fetch-increment and store-add are examples of communications register instructions. Further there is an inter-CPU interrupt instruction as well as a multi-CPU interrupt instruction. The interrupt instructions are useful for scheduling and synchronization as well as debugging support.

There is a second level global communications register set in the IXS.

---

[1] Cray is a registered trademark of Cray Inc. All other trademarks are the property of their respective owners

[2] IBM is a registered trademark of International Business Machines Corporation or its wholly owned subsidiaries

3.11 Memory Unit

To achieve efficient vector processing a large main memory and high memory throughput that match the processor performance are required. Whereas the SX-6 supported SDRAM only the SX-8 supports both DDR2-SDRAM as well as fast cycle memory (FCRAM). Fast cycle memory provides faster access and lower latency compared to DDR2-SDRAM, however with DDR2-SDRAM larger capacities can be realized. With fast cycle memory up to 64 GB main memory can be supported within a single 8 CPU node, with DDR2-SDRAM 128 GB. For the FCRAM memory type the bandwidth between each CPU and the main memory is 64 GB/s thus realizing an aggregated memory throughput of 512 GB/s within a single node and for the DDR2-SDRAM memory type the bandwidth between each CPU and the main memory is 70.4 GB/s thus realizing an aggregated memory throughput of 563.2 GB/s within a single node.

The memory architecture within each single-node frame is a non-blocking crossbar that provides uniform high-speed access to the main memory. This constitutes a symmetric multiprocessor shared memory system (SMP) also known as a parallel vector processor (PVP).

SX-8 Series systems are real memory mode machines but utilize page mapped addressing. Demand paging is not supported. The page mapped architecture allows load modules to be non-contiguously loaded, eliminating the need for periodic memory compaction procedures by the operating system and enabling the most efficient operational management techniques possible. Another advantage of the page mapped architecture is that in case of swapping only that number of pages needs to be swapped out which are needed to swap another job in thus reducing I/O wait time considerably.

The processor to memory port is classified as a single port per processor. Either load or store can occur during any transfer cycle. Each SX processor automatically reorders main memory requests in two important ways. Memory references look-ahead and pre-issue are performed to maximize throughput and minimize memory waits. The issue unit reorders load and store operations to maximize memory path efficiency.

The availability of the programmable vector data registers significantly reduces memory traffic as compared to a system without programmable vector data registers. Because of the programmable vector data registers in the general case an SX-8 Series system requires only 50–60% of the memory bandwidth required by traditional architectures which only use vector operational registers. Consider also that the bandwidth available to the SX-8 Series processor is sufficient for its peak performance rating and is substantially higher than other competing systems.

3.12 Input-Output Feature (IOF)

Each SX-8 node can have up to 4 I/O features (IOF) which provide for an aggregate I/O bandwidth of 12.8 GB/s. The IOF can be equipped with up to

55 channel cards which support industry standard interfaces such as 2 Gb FC, Ultra320-SCSI, 1000base-SX, 10/100/1000base-T. Support for 4 Gb and 10 Gb FC, 10 Gb Ethernet and others are planned.

The IOFs operate asynchronously with the processors as independent I/O engines so that central processors are not directly involved in reading and writing to storage media as is the case of workstation technology based systems.

To further offload the CPU from slow I/O operations a new I/O architecture is being introduced in SX-8. The conventional firmware interface between memory and host bus adapters has been replaced by direct I/O between memory and intelligent host bus adapters.

3.13 FC Channels

The SX-8 Series offers native FC channels (2 Gbps) for the connection of the latest, highly reliable, high performance peripheral devices such as RAID disks. FC offers the advantages of connectivity to newer high performance RAID storage systems that are approaching commodity price levels. Further, numerous storage devices can be connected to FC.

3.14 SCSI Channels

Ultra320-SCSI channels are available. Direct SCSI support enables the configuration of very low cost commodity storage devices when capacity outweighs performance criteria.

A large component of SCSI connected disk is not recommended because of the performance mismatch to the SX-8 Series system. FC storage devices are highly recommended to maintain the necessary I/O rates for the SX-8 Series.

SCSI channels are most useful for connecting to tape devices. Most tape devices easily maintain their maximum data rate via the SCSI channel interfaces.

3.15 Internode Crossbar Switch (IXS)

The IXS (Inter-node Crossbar Switch) is a NEC proprietary device that connects SX-8 nodes in a highly efficient way.

Each SX-8 node is equipped with 2 Remote Control Units (RCU) that connect the SX-8 to the IXS. Utilizing the two RCUs allow for a maximum of 512 SX-8 nodes to be connected to a single IXS with a bandwidth of 32 GB/s per node.

The IXS is a full crossbar providing a high speed single stage non-blocking interconnect with an aggregate bi-directional bandwidth of 16 TB/s. IXS facilities provided include inter-node addressing and page mapping, remote unit control, inter-node data movement, and remote processor instruction support (e.g., interrupt of a remote CPU). It also contains system global communications registers to enable efficient software synchronization of events occurring across multiple nodes. There are 8 × 64 bit global communications registers available for each node.

Both synchronous and asynchronous transfers are supported. Synchronous transfers are limited to 2 kB, and asynchronous transfers to 32 MB. This is transparent to the user as it is entirely controlled by the NEC MPI library.

The interface technology is based on 3 Gbps (gigabits per second) optical interfaces providing approximately 2.7 µs (microsecond) node-to-node hardware latency (with a 20 m cable length) and 16 GB/s of node-to-node bi-directional bandwidth.

The minimum (best effort) time for a broadcast to reach all nodes in a 512 node configuration would be:

$$\text{(latency of node-to-node)} \cdot \log_2 \text{(node count)} = 24.3\,\mu\text{s}$$

The two RCUs allow the following three types of SX-8 Multi Node Systems to be configured:

- 512 Nodes connected to a single IXS with a bandwidth of bidirectional 16 GB/s per node)
- 256 SX-8 Nodes connected to a single IXS with a bandwidth of bidirectional 32 GB/s per node (Fig. 6)
- 512 SX-8 Nodes connected to two IXS switches with a bandwidth of bidirectional 16 GB/s per node as a fail-safe configuration.

The IXS provides very tight coupling between nodes virtually enabling a single system image both from a hardware and a software point of view.

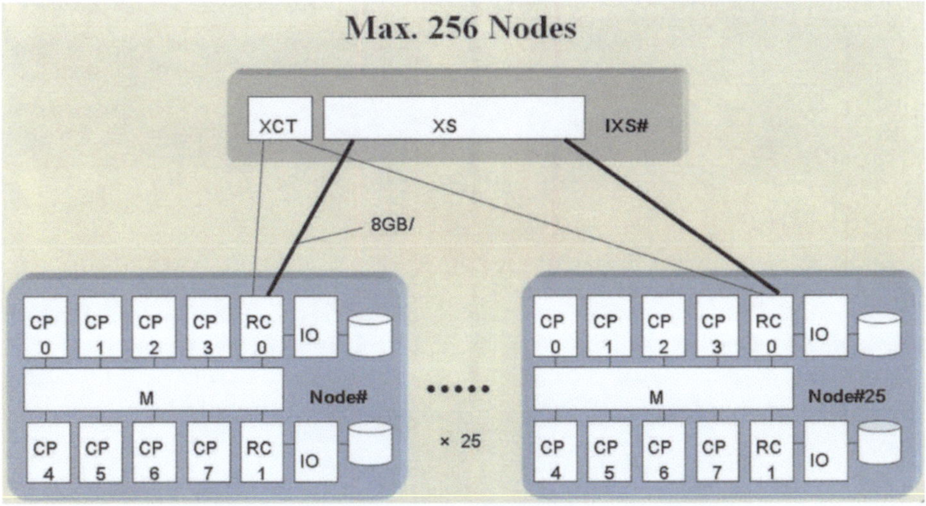

Fig. 6. Single IXS connection, max. 256 nodes

4 Software

4.1 Operating System

The SX Series SUPER-UX operating system is a System V port with additional features from 4.3 BSD plus enhancements to support supercomputing requirements (Fig. 7). It has been in widespread production use since 1990. Through the inheritance of the SX architecture, the operating system SUPER-UX maintains perfect compatibility with the SX-6/SX-5 Series.

Some recent major enhancements include:

- Enhancing the Multi Node system
  - Enhancement of the ERSII (Enhanced Resource Scheduler II) enabling reflection of site-specific policy to job scheduling in order to support Multi Node systems
  - Association (sharing files) with the IA64/Linux server using GFS (NECs Global File System) which provides high-speed inter-node file sharing.
- Support for Fortran, C and C++ cross-compilers that run on Linux and various other platforms
  - Maturity of Etnus TotalView port, including enhanced functionality and performance in controlling multi-tasked programs, cleaner display of C++ (including template) F90 types, etc.

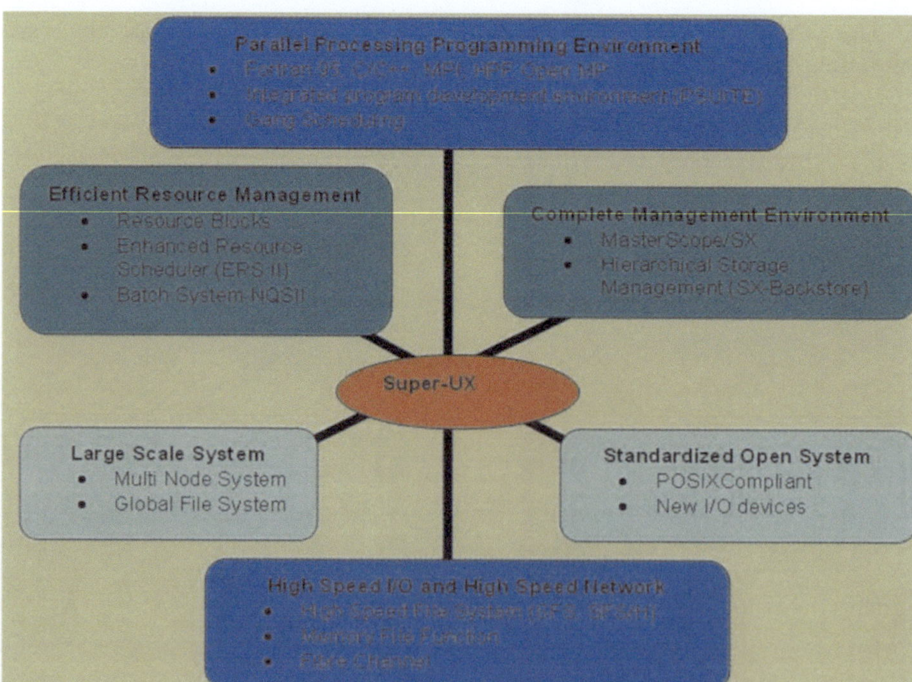

Fig. 7. Super UX Features

  - Enhanced features of Vampir/SX allowing easier analysis of programs that runs on large-scale Multi Node SX systems
- Operation improvements
  - Enhancement of MasterScope/SX which simplifies the overall management of systems integrated over a network. This allows the operation of a Multi Node system under a single system image.
  - Enhancement of the batch system NQSII

Automatic Operation

SX systems provide hardware and software support options to enable operatorless environments. The system can be pre-programmed to power on, boot, enter multi-user mode and shutdown/power-off under any number of programmable scenarios. Any event that can be determined by software and responded to by closing a relay or executing a script can be serviced.

The automatic operation system includes a hardware device called the Automatic Operation Controller (AOC). The AOC serves as an external control and monitoring device for the SX system. The AOC can perform total environmental monitoring, including earthquake detection.

Cooperating software executing in the SX system communicates system load status and enables the automatic operation system to execute all UNIX functions necessary for system operation.

NQSII Batch Subsystem

SUPER-UX NQSII (Network Queuing System II) is a batch processing system for the maximum utilization of high-performance cluster system computing resources. The NQSII enhances system operability by monitoring workloads of the computing nodes that comprise the cluster system to achieve load sharing of the entire cluster system by reinforcing the single system image (SSI) and providing a single system environment (SSE). The functionalities of the NQSII are reinforced by tailoring the major functions including the job queuing, resource management and load balancing to the cluster system while implementing conventional NQS features. Load balancing is further enhanced by using an extended scheduler (ERSII) tailored to the NQSII. Also, a fair share scheduling mechanism can be utilized.

NQSII is enhanced to add substantial user control over work in progress. File Staging transfers the files which relate to the execution of batch job among the client host and the execution hosts.

NQSII queues and the full range of individual queue parameters and accounting facilities are supported. The NQSII queues have substantial scheduling parameters available including time slices, maximum cpu time, maximum memory sizes, etc.

NQSII batch request can be checkpointed by the owner, operator, or NQSII administrator. No special programming is required for checkpointing.

Checkpoint/restart is valuable for interrupting very long executions for preventative maintenance or to provide a restart mechanism in case of catastrophic system failure or for recovery of correctable data errors.

NQSII batch jobs can be job migrated. The process to move a job managed by a job server to the control of another job server is called job migration; this can be used to equalize the load on the executing hosts (nodes).

Enhanced Configuration Options and Logical Partitioning

SUPER-UX has a feature called Resource Block Control that allows the system administrator to define logical scheduling groups that are mapped onto the SX-8 processors.

Each Resource Block has a maximum and minimum processor count, memory limits and scheduling characteristics such that the SX-8 can be defined as multiple logical environments. For example, one portion of an SX-8 can be defined primarily for interactive work while another may be designated for non-swappable parallel processing scheduling using a FIFO scheme and a third area can be configured to optimize a traditional parallel vector batch environment. In each case any resources not used can be "borrowed" by other Resource Blocks.

Supercomputer File System

The SUPER-UX native file system is called SFS. It has a flexible file system level caching scheme utilizing XMU space; numerous parameters can be set including cache size, threshold limits and allocation cluster size. Files can be 512 TB in size because of 64 bit pointer utilization.

SFS has a number of advanced features including methods to handle both device overflow and file systems that span multiple physical devices.

- Supercomputer File System, Special Performance
  The special performance Supercomputer File System (SFS/H) has reduced overhead, even compared to the highly efficient SFS.
  Limitations in use of SFS/H files are that transfers must be in 4-byte multiples. This restriction is commonly met in FORTRAN but general case UNIX programs often write arbitrary length byte streams which must be directed to an SFS file system.
- Global File System
  SUPER-UX Multi Node systems have a Global File System (GFS) that enables the entire Multi Node complex to view a single coherent file system. GFS works as a client-server concept. NEC has implemented the GFS server function on its IA64 based server (TX7). The server manages the I/O requests from the individual clients. The actual I/O however is being executed directly between the global disk subsystem and the requesting clients. Clients are not only available for NEC products like SX-8 and TX7 but are or will soon become available also for various other popular server platforms like HP (HP-UX), IBM (AIX), SGI (IPF Linux), SUN (SOLARIS) as well as on PC Clusters (IA32-LINUX).

– MFF Memory File Facility
  The SX-MFF (Memory File Facility) is available to enable high performance I/O caching features and to provide very high performance file system area that is resident in main memory.

Multilevel Security

The Multilevel Security (MLS) option is provided to support site requirements for either classified projects or restricted and controlled access. Security levels are site definable as to both names and relationships. MLS has been in production use since early 1994.

Multi Node Facility

SX-8/M Multi Node systems provide special features to enable efficient use of the total system.

The Global File System provides a global file directory for all components of the node and for all jobs executing anywhere on the Multi Node system.

The SUPER-UX kernel is enhanced to recognize a Multi Node job class. When a Multi Node job (i.e., a job using processors on more than one node) enters the system, the kernel properly sequences all of the master processes across the nodes, initializes the IXS Super-Switch translation pages for the job and provides specialized scheduling commensurate with the resources being used. Once initialization is complete the distributed processes can communicate with each other without further operating system involvement.

RAS Features

In the SX-8 Series a dramatic improvement in hardware reliability is realized by using the latest technology and high-integration designs such as the single-chip vector processor while further reducing the number of parts. As with conventional machines, error-correcting codes in main memory and error detecting functions such as circuit duplication and parity checks have been implemented in the SX-8. When a hardware error does occur a built-in diagnostics function (BID) quickly and automatically indicates the location of the fault and an automatic reconfiguration function releases the faulty component and continues system operation. In addition to the functions above, prompt fault diagnosis and simplified preventive maintenance procedures using automatic collection of fault information, automatic reporting to the service center and remote maintenance from the service center result in a comprehensive improvement in the systems reliability, availability and serviceability.

4.2 Compilers

Fortran95, C, C++ and HPF languages are supported on the SX-8. An optimized MPI library both supporting the complete MPI-1 and MPI-2 standard is available on SX-8.

The compilers provide advanced automatic optimization features including automatic vectorization and automatic parallelization, partial and conditional vectorization, index migration, loop collapsing, nested loop vectorization, conversions, common expression elimination, code motion, exponentiation optimization, optimization of masked operations, loop unrolling, loop fusion, inline subroutine expansion, conversion of division to multiplication and instruction scheduling.

Compiler options and directives provide the programmer with considerable flexibility and control of compilation and optimizations.

FORTRAN90/SX

FORTRAN90/SX is offered as a native compiler as well as a workstation based cross development system that includes full compile and link functionality. FORTRAN90/SX offers automatic vectorization and parallelization applied to standard portable Fortran95 codes. In addition to the listed advanced optimization features, FORTRAN90/SX includes data trace analysis and a performance data feedback facility. FORTRAN90/SX also supports OpenMP 2.0 and Microtasking.

HPF/SX

HPF development on SUPER-UX is targeted toward SX Series Multi Node systems. NEC participates in the various HPF forums working in the United States and Japan with the goal of further developing and improving the HPF language. HPF2 is underway and there is an HPF Japan forum (HPFJA) that is sponsoring additional enhancements to the HPF2 language.

C++/SX

C++/SX includes both C and C++ compilers. They share the "back end" with FORTRAN90/SX and as such provide comparable automatic vectorization and parallelization features. Additionally they have rich optimization features of pointer and structure operations.

4.3 Programming Tools

PSUITE Integrated Program Development Environment

PSUITE is the integrated program development environment for SUPER-UX. It is available as a cross environment for most popular workstations. It operates cooperatively with the network-connected SX-8 system. PSUITE supports FORTRAN90/SX and C++/SX applications development.

Editing and Compiling

The built-in Source Browser enables the user to edit source programs. For compiling all major compiler options are available through pull downs and X-Window style boxes. Commonly used options can be enabled with buttons and free format boxes are available to enter specific strings for compilation and linking. Figure 8 shows the integration of the compiler options windows with the Source Browser.

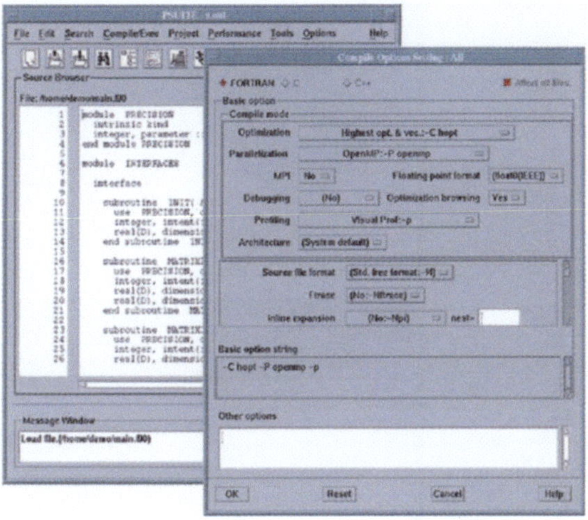

Fig. 8. Compiler Option Window with Source Browser

Debugging

Debugging is accomplished through the PDBX. PDBX being the symbolic debugger for shared memory parallel programs. Enhanced capabilities include the graphical presentation of data arrays in various 2 or 3 dimensional styles.

Application Tuning

PSUITE has two performance measurement tools. One is Visual Prof which measures performance information easily. The other one is PSUITEperf which measures performance information in detail. By analyzing the performance using them the user can locate the program area in which a performance problem lies. Correcting these problems can improve the program performance. Figure 9 shows performance information measured by PSUITEperf.

4.4 FSA/SX

FSA/SX is a static analysis tool that outputs useful analytical information for tuning and porting of programs written in FORTRAN. It can be used with either a command line interface or GUI.

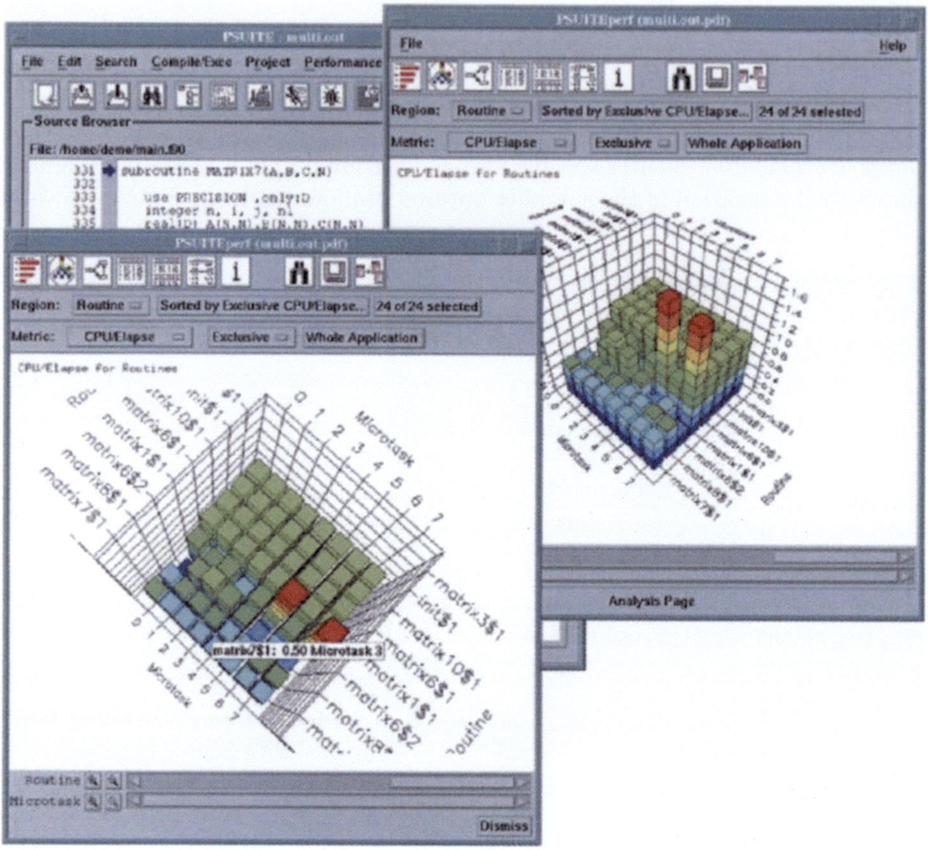

Fig. 9. PSuite Performance View

4.5 TotalView

TotalView is the debugger provided by Etnus which has been very popular for use on HPC platforms including the SX. TotalView for SX-8 system supports FORTRAN90/SX, C++/SX programs and MPI/SX programs. The various functionalities of TotalView enable easy and efficient development of complicated parallel and distributed applications.

Figure 10 shows the process window, the call-tree window and Message queue graph window. The process window in the background shows source code, stack trace (upper-left), stack frame (upper-right) for one or more threads in the selected process. The message queue graph window on the right hand side shows MPI program's message queue state of the selected communicator graphically. The call-tree window (at the bottom) shows a diagram linking all the currently-active routines in all the processes or the selected process by arrows annotated with calling frequency of one routine by another.

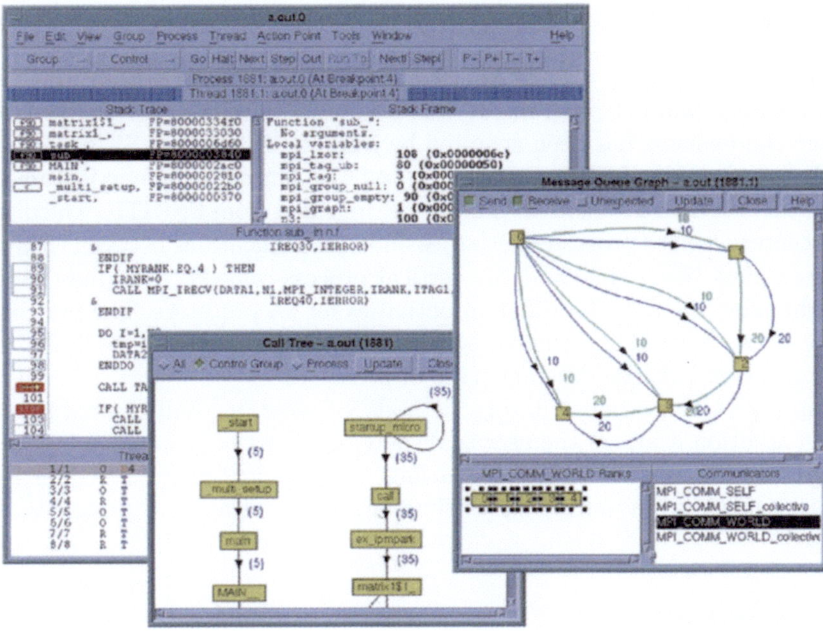

Fig. 10. TotalView

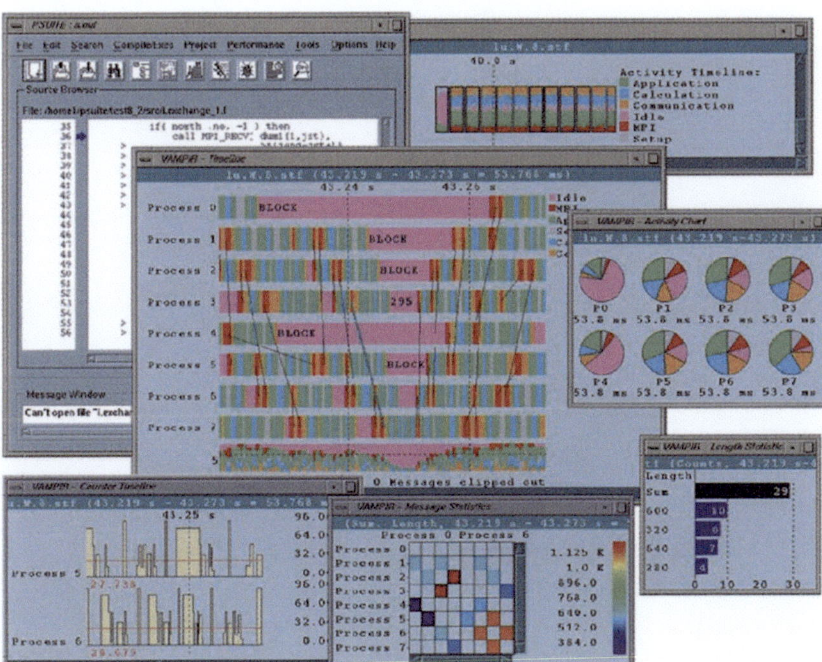

Fig. 11. Vampir/SX

4.6 Vampir/SX

Vampir/SX enables the user to examine execution characteristics of the distributed-memory parallel program. It was originally developed by Pallas GmbH (though the business has been acquired by Intel) and ported to SX series. Vampir/SX has all major features of Vampir and also has some unique features.

Figure 11 shows a session of Vampir/SX initiated from PSUITE. The display in the center outlines processes activities and communications between them, the horizontal axis being time and the vertical process-rank(id). The pie charts to the right show the ratio for different activities for all processes. The matrix-like display at the bottom and the bar-graph to the bottom-right shows statistics of communication between different pairs of processes.

Vampir/SX has various filtering methods for recording only desired information. In addition it allows the user to display only part of recorded information, saving time and memory used for drawing. The window to the top-right is the interface allowing the user to select time-intervals and a set of processes to be analyzed.

4.7 Networking

All normal UNIX communications protocols are supported. SUPER-UX supports Network File System (NFS) Versions 2 and 3.

Have the Vectors the Continuing Ability to Parry the Attack of the Killer Micros?

Peter Lammers[1], Gerhard Wellein[2], Thomas Zeiser[2], Georg Hager[2], and Michael Breuer[3]

[1] High Performance Computing Center Stuttgart (HLRS), Nobelstraße 19, D-70569 Stuttgart, Germany, plammers@hlrs.de,
[2] Regionales Rechenzentrum Erlangen (RRZE), Martensstraße 1, D-91058 Erlangen, Germany, hpc@rrze.uni-erlangen.de,
[3] Institute of Fluid Mechanics (LSTM), Cauerstraße 4, D-91058 Erlangen, Germany, breuer@lstm.uni-erlangen.de

Abstract Classical vector systems still combine excellent performance with a well established optimization approach. On the other hand clusters based on commodity microprocessors offer comparable peak performance at very low costs. In the context of the introduction of the NEC SX-8 vector computer series we compare single and parallel performance of two CFD (computational fluid dynamics) applications on the SX-8 and on the SGI Altix architecture demonstrating the potential of the SX-8 for teraflop computing in the area of turbulence research for incompressible fluids. The two codes use either a finite-volume discretization or implement a lattice Boltzmann approach, respectively.

1 Introduction

Starting with the famous talk of Eugene Brooks at SC 1989 [1] there has been an intense discussion about the future of vector computers for more than 15 years. Less than 5 years ago, right at the time when it was widely believed in the community that the "killer micros" have finally succeeded, the "vectors" stroke back with the installation of the NEC Earth Simulator (ES). Furthermore, the U.S. re-entered vector territory, allowing CRAY to go back to its roots.

Even though massively parallel systems or clusters based on microprocessors deliver high peak performance and large amounts of compute cycles at a very low price tag, it has been emphasized recently that vector technology is still extremely competitive or even superior to the "killer micros" if application performance for memory intensive codes is the yardstick [2, 3, 4].

Introducing the new NEC SX-8 series in 2005, the powerful technology used in the ES has been pushed to new performance levels by doubling all important performance metrics like peak performance, memory bandwidth and interconnect bandwidth. Since the basic architecture of the system itself did not change at all from a programmer's point of view, the new system is expected to run most applications roughly twice as fast as its predecessor, even using the same binary.

In this report we test the potentials of the new NEC SX-8 architecture using selected real world applications from CFD and compare the results with the predecessor system (NEC SX-6+) as well as a microprocessor based system. For the latter we have chosen the SGI Altix, which uses Intel Itanium 2 processors and usually provides high efficiencies for the applications under consideration in this report.

We focus on two CFD codes from turbulence research, both being members of the HLRS TERAFLOP-Workbench [5], namely DIMPLE and TeraBEST. The first one is a classical finite-volume code called $\mathcal{LESOCC}$ (Large Eddy Simulation On Curvilinear Co-ordinates [6, 7, 8, 9]), mainly written in FORTRAN77. The second one is a more recent lattice Boltzmann solver called $\mathcal{BEST}$ (Boltzmann Equation Solver Tool [10]) written in FORTRAN90. Both codes are MPI-parallelized using domain decomposition and have been optimized for a wide range of computer architectures (see e.g. [11, 12]). As a test case we run simulations of flow in a long plane channel with square cross section or over a single flat plate. These flow problems are intensively studied in the context of wall-bounded turbulence.

2 Architectural Specifications

From a programmer's view, the NEC SX-8 is a traditional vector processor with 4-track vector pipes running at 2 GHz. One multiply and one add instruction per cycle can be sustained by the arithmetic pipes, delivering a theoretical peak performance of 16 GFlop/s. The memory bandwidth of 64 GByte/s allows for one load or store per multiply-add instruction, providing a balance of 0.5 Word/Flop. The processor has 64 vector registers, each holding 256 64-bit words. Basic changes compared to its predecessor systems are a separate hardware square root/divide unit and a "memory cache" which lifts stride-2 memory access patterns to the same performance as contiguous memory access. An SMP node comprises eight processors and provides a total memory bandwidth of 512 GByte/s, i. e. the aggregated single processor bandwidths can be saturated. The SX-8 nodes are networked by an interconnect called IXS, providing a bidirectional bandwidth of 16 GByte/s and a latency of about 5 microseconds.

For a comparison with the technology used in the ES we have chosen a NEC SX-6+ system which implements the same processor technology as used in the ES but runs at a clock speed of 565 MHz instead of 500 MHz. In contrast to the NEC SX-8 this vector processor generation is still equipped with two 8-track vector pipelines allowing for a peak performance of 9.04 GFlop/s per CPU for the NEC SX-6+ system. Note that the balance between main memory bandwidth and peak performance is the same as for the SX-8 (0.5 Word/Flop) both for the single processor and the 8-way SMP node. Thus, we expect most application codes to achieve a speed-up of around 1.77 when going from SX-6+ to SX-8. Due to the architectural changes described above the SX-8 should be able to show even a better speed-up on some selected codes.

As a competitor we have chosen the SGI Altix architecture which is based on the Intel Itanium 2 processor. This CPU has a superscalar 64-bit architecture providing two multiply-add units and uses the Explicitly Parallel Instruction Computing (EPIC) paradigm. Contrary to traditional scalar processors, there is no out-of-order execution. Instead, compilers are required to identify and exploit instruction level parallelism. Today clock frequencies of up to 1.6 GHz and on-chip caches with up to 9 MBytes are available. The basic building block of the Altix is a 2-way SMP node offering 6.4 GByte/s memory bandwidth to both CPUs, i.e. a balance of 0.06 Word/Flop per CPU. The SGI Altix3700Bx2 (SGI Altix3700) architecture as used for the $\mathcal{BEST}$ ($\mathcal{LESOCC}$) application is based on the NUMALink4 (NUMALink3) interconnect, which provides up to 3.2 (1.6) GByte/s bidirectional interconnect bandwidth between any two nodes and latencies as low as 2 microseconds. The NUMALink technology allows to build up large powerful shared memory nodes with up to 512 CPUs running a single Linux OS.

The benchmark results presented in this paper were measured on the NEC SX-8 system (576 CPUs) at High Performance Computing Center Stuttgart (HLRS), the SGI Altix3700Bx2 (128 CPUs, 1.6 GHz/6 MB L3) at Leibniz Rechenzentrum München (LRZ) and the SGI Altix3700 (128 CPUs, 1.5 GHz/6 MB L3) at CSAR Manchester.

All performance numbers are given either in GFlop/s or, especially for the lattice Boltzmann application, in MLup/s (**M**ega **L**attice Site **U**pdates **p**er **S**econd), which is a handy unit for measuring the performance of LBM.

3 Finite-Volume-Code $\mathcal{LESOCC}$

3.1 Background and Implementation

The CFD code $\mathcal{LESOCC}$ was developed for the simulation of complex turbulent flows using either the methodology of direct numerical simulation (DNS), large-eddy simulation (LES), or hybrid LES-RANS coupling such as the detached-eddy simulation (DES).

$\mathcal{LESOCC}$ is based on a 3-D finite-volume method for arbitrary non-orthogonal and non-staggered, block-structured grids [6, 7, 8, 9]. The spatial discretization of all fluxes is based on central differences of second-order accuracy. A low-storage multi-stage Runge-Kutta method (second-order accurate) is applied for time-marching. In order to ensure the coupling of pressure and velocity fields on non-staggered grids, the momentum interpolation technique is used. For modeling the non-resolvable subgrid scales, a variety of different models is implemented, cf. the well-known Smagorinsky model [13] with Van Driest damping near solid walls and the dynamic approach [14, 15] with a Smagorinsky base model.

$\mathcal{LESOCC}$ is highly vectorized and additionally parallelized by domain decomposition using MPI. The block structure builds the natural basis for grid partitioning. If required, the geometric block structure can be further subdivided into a parallel block structure in order to distribute the computational load to a number of processors (or nodes).

Because the code was originally developed for high-performance vector computers such as CRAY, NEC or Fujitsu, it achieves high vectorization ratios ($> 99.8\%$). In the context of vectorization, three different types of loop structures have to be distinguished:

- Loops running linearly over all internal control volumes in a grid block (3-D volume data) and exhibit no data dependencies. These loops are easy to vectorize, their loop length is much larger than the length of the vector registers and they run at high performance on all vector architectures. They show up in large parts of the code, e.g. in the calculation of the coefficients and source terms of the linearized conservation equations.
- The second class of loops occurs in the calculation of boundary conditions. Owing to the restriction to 2-D surface data, the vector length is shorter than for the first type of loops. However, no data dependence prevents the vectorization of this part of the code.
- The most complicated loop structure occurs in the solver for the linear systems of equations in the implicit part of the code. Presently, we use the strongly implicit procedure (SIP) of Stone [16], a variant of the incomplete LU (ILU) factorization. All ILU type solvers of standard form are affected by recursive references to matrix elements which would in general prevent vectorization. However, a well-known remedy for this problem exists. First, we have to introduce diagonal planes (hyper-planes) defined by $i + j + k =$ constant, where i, j, and k are the grid indices. Based on these hyper-planes we can decompose the solution procedure for the whole domain into one loop over all control volumes in a hyper-plane where the solution is dependent only on the values computed in the previous hyper-plane and an outer do-loop over the $i_{\max} + j_{\max} + k_{\max} - 8$ hyper-planes.

3.2 Performance of $\mathcal{LESOCC}$

The most time-consuming part of the solution procedure is usually the implementation of the incompressibility constraint. Profiling reveals that $\mathcal{LESOCC}$ spends typically 20–60% of the total runtime in the SIP-solver, depending on the actual flow problem and computer architecture. For that reason we have established a benchmark kernel for the SIP-solver called SipBench [17], which contains the performance characteristics of the solver routine and is easy to analyze and modify. In order to test for memory bandwidth restrictions we have also added an OpenMP parallelization to the different architecture-specific implementations.

In Fig. 1 we show performance numbers for the NEC SX-8 using a hyperplane implementation together with the performance of the SGI Altix which uses a pipeline-parallel implementation (cf. [11]) on up to 16 threads. On both

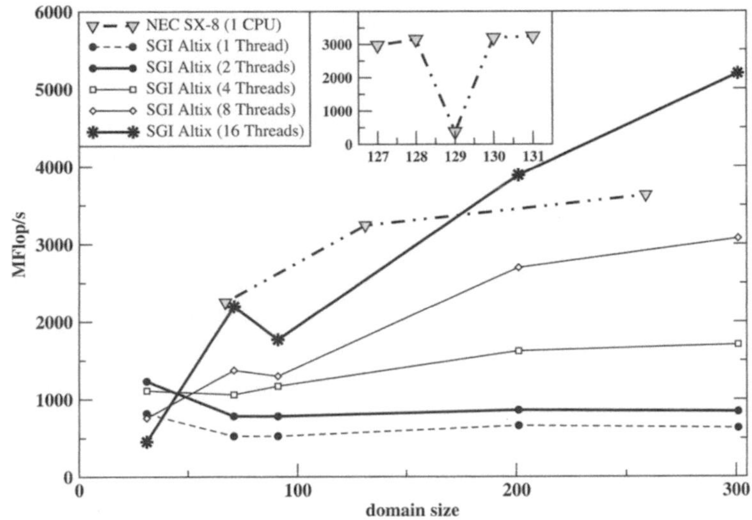

Fig. 1. Performance of SipBench for different (cubic) domains on SGI Altix using up to 16 threads and on NEC SX-8 (single CPU performance only)

machines we observe start-up effects (vector pipeline or thread synchronisation), yielding low performance on small domains and saturation at high performance on large domains. For the pipeline-parallel (SGI Altix) 3-D implementation a maximum performance of 1 GFlop/s can be estimated theoretically, if we assume that the available memory bandwidth of 6.4 GByte/s is the limiting factor and caches can hold at least two planes of the 3D domain for the residual vector. Since two threads (sharing a single bus with 6.4 GByte/s bandwidth) come very close (800 MFlop/s) to this limit we assume that our implementation is reasonably optimized and pipelining as well as latency effects need not be further investigated for this report.

For the NEC SX-8 we use a hyper-plane implementation of the SIP-solver. Compared to the 3-D implementation additional data transfer from main memory and indirect addressing is required. Ignoring the latter, a maximum performance of 6–7 GFlop/s can be expected on the NEC SX-8. As can be seen from Fig. 1, with a performance of roughly 3.5 GFlop/s the NEC system falls short of this expectation. Removing the indirect addressing one can achieve up to 5 GFlop/s, however at the cost of substantially lower performance for small/intermediate domain sizes or non-cubic domains. Since this is the application regime for our $\mathcal{LESOCC}$ benchmark scenario we do not discuss the latter version in this report. The inset of Fig. 1 shows the performance impact of slight changes in domain size. It reveals that solver performance can drop by a factor of 10 for specific memory access patterns, indicating severe memory bank conflicts.

The other parts of $\mathcal{LESOCC}$ perform significantly better, liftig the total single processor performance for a cubic plane channel flow scenario with 130^3 grid points to 8.2 GFlop/s on the SX-8. Using the same executable we measured a performance of 4.8 GFlop/s GFlop/s on a single NEC SX-6+ processor, i.e. the SX-8 provides a speedup of 1.71 which is in line with our expectations based on the pure hardware numbers.

For our strong scaling parallel benchmark measurements we have chosen a boundary layer flow over a flat plate with 11×10^6 grid points and focus on moderate CPU counts (6, 12 and 24 CPUs), where the domain decomposition for $\mathcal{LESOCC}$ can be reasonably done. For the 6 CPU run the domain was cut in wall-normal direction only; at 12 and 24 CPUs streamwise cuts have been introduced, lowering the communication-to-computation ratio.

The absolute parallel performance for the NEC SX-8 and the SGI Altix systems is depicted in Fig. 2. The parallel speedup on the NEC machine is obviously not as perfect as on the Altix system. Mainly two effects are responsible for this behavior. First, the baseline measurements with 6 CPUs were done in a single node on the NEC machine ignoring the effect of communication over the IXS. Second, but probably more important the single CPU performance (cf. Table 1) of the vector machine is almost an order of magnitude higher than on the Itanium 2 based system, which substantially increases the impact of communication on total performance due to strong scaling. A more detailed profiling of the code further reveals that also the performance of the SIP-solver is reduced with increasing CPU count on the NEC machine due to reduced vector length (i.e. smaller domain size per CPU).

The single CPU performance ratio between vector machine and cache based architecture is between 7 and 9.6. Note that we achieve a L3 cache hit ratio of roughly 97% (i.e. each data element loaded from main memory to cache can be

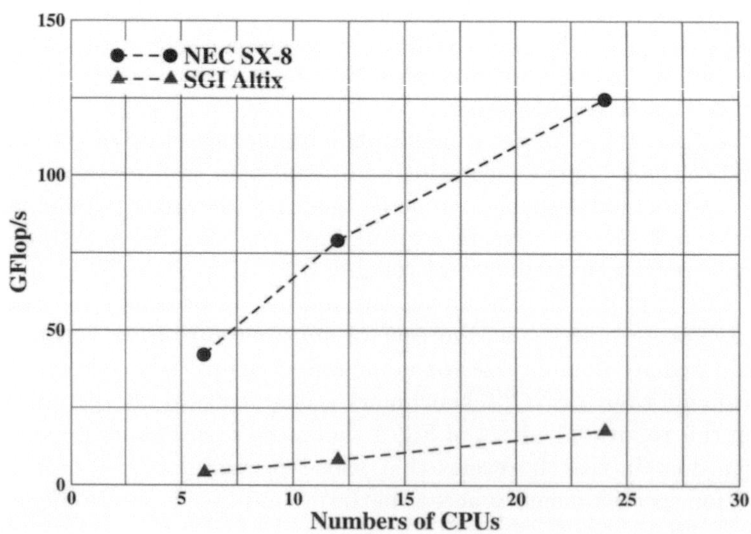

Fig. 2. Speedup (strong scaling) for a boundary layer flow with 11×10^6 grid points up to 24 CPUs

Table 1. Fraction of the SIP-solver and its performance in comparison of the overall performance. Data from the boundary layer setup with 24 CPUs

Platform	**CPUs**	**Time SIP-solver** (%)	**SIP-solver** GFlop/s/CPU	**L3 cache-hit rate** (%)	*LESOCC* GFlop/s/CPU
Intel Itanium 2 (1.6 GHz)	24	25	0.39	97	0.73
NEC SX-8 (2 GHz)	6	31.5	3.25	—	7.02
NEC SX-8 (2 GHz)	12	32.9	2.83	—	6.6
NEC SX-8 (2 GHz)	24	33.6	2.5	—	5.2

reused at least once from cache), which is substantially higher than for purely memory bound applications.

4 Lattice Boltzmann Code *BEST*

4.1 Background and Implementation

The original motivation for the development of *BEST* was the ability of the lattice Boltzmann method to handle flows through highly complex geometries very accurately and efficiently. This refers not only to the flow simulation itself but also to the grid generation which can be done quite easily by using the "marker and cell" approach. Applying the method also to the field of numerical simulation (DNS or LES) of turbulence might be further justified by the comparatively very low effort per grid point. In comparison to spectral methods the effort is lower at least by a factor of five [10]. Furthermore, the method is based on highly structured grids which is a big advantage for exploiting all kinds of hardware architectures efficiently. On the other hand this might imply much larger grids than normally used by classical methods.

The widely used class of lattice Boltzmann models with BGK approximation of the collision process [18, 19, 20] is based on the evolution equation

$$f_i(\mathbf{x} + \mathbf{e}_i \delta t,\, t + \delta t) = f_i(\mathbf{x},\, t) - \frac{1}{\tau}\left[f_i(\mathbf{x},\, t) - f_i^{\mathrm{eq}}(\rho, \mathbf{u})\right], \qquad i = 0 \ldots N\,. \tag{1}$$

Here, f_i denotes the particle distribution function which represents the fraction of particles located in timestep t at position $\mathbf{x}$ and moving with the microscopic velocity $\mathbf{e}_i$. The relaxation time τ determines the rate of approach to local equilibrium and is related to the kinematic viscosity of the fluid. The equilibrium state f_i^{eq} itself is a low Mach number approximation of the Maxwell-Boltzmann equilibrium distribution function. It depends only on the macroscopic values of the fluid density ρ and the flow velocity $\mathbf{u}$. Both can be easily obtained as first moments of the particle distribution function.

The discrete velocity vectors $\mathbf{e}_i$ arise from the N chosen collocation points of the velocity-discrete Boltzmann equation and determine the basic structure of the numerical grid. We choose the D3Q19 model [18] for discretization in 3-D, which uses 19 discrete velocities (collocation points) and provides a computational domain with equidistant Cartesian cells (voxels).

Each timestep ($t \rightarrow t + \delta t$) consists of the following steps which are repeated for all cells:

- Calculation of the local macroscopic flow quantities ρ and $\mathbf{u}$ from the distribution functions, $\rho = \sum_{i=0}^{N} f_i$ and $\mathbf{u} = \frac{1}{\rho} \sum_{i=0}^{N} f_i \mathbf{e}_i$.
- Calculation of the equilibrium distribution f_i^{eq} from the macroscopic flow quantities (see [18] for the equation and parameters) and execution of the "collision" (relaxation) process, $f_i^*(\mathbf{x}, t^*) = f_i(\mathbf{x}, t) - \frac{1}{\tau} [f_i(\mathbf{x}, t) - f_i^{\mathrm{eq}}(\rho, \mathbf{u})]$, where the superscript * denotes the post-collision state.
- "Propagation" of the $i = 0 \ldots N$ post-collision states $f_i^*(\mathbf{x}, t^*)$ to the appropriate neighboring cells according to the direction of $\mathbf{e}_i$, resulting in $f_i(\mathbf{x} + \mathbf{e}_i \delta t, t + \delta t)$, i.e. the values of the next timestep.

The first two steps are computationally intensive but involve only values of the local node while the third step is just a direction-dependent uniform shift of data in memory. A fourth step, the so called "bounce back" rule [19, 20], is incorporated as an additional part of the propagation step and "reflects" the distribution functions at the interface between fluid and solid cells, resulting in an approximate no-slip boundary condition at walls.

Of course the code has to be vectorized for the SX. This can easily be done by using two arrays for the successive time steps. Additionally, the collision and the propagation step are collapsed in one loop. This reduces transfers to main memory within one time step. Consequently $B = 2 \times 19 \times 8$ Bytes have to be transferred per lattice site update. The collision itself involves roughly $F = 200$ floating point operations per lattice site update. Hence one can estimate the achievable theoretical peak performance from the basic performance characteristics of an architecture such as memory bandwidth and peak performance. If performance is limited by memory bandwidth, this is given by $P = \mathrm{MemBW/B}$ or by $P = \mathrm{PeakPerf/F}$ if it is limited by the peak performance.

4.2 Performance of $\mathcal{BEST}$

The performance limits imposed by the hardware together with the measured performance value can be found in Table 2. Whereas Itanium 2 is clearly limited by its memory bandwidth, the SX-8 ironically suffers from its "low peak performance". This is true for the NEC SX-6+ as well.

In Fig. 3 single and parallel performance of $\mathcal{BEST}$ on the NEC SX-8 are documented. First of all, the single CPU performance is viewed in more detail regarding the influence of vector length. The curve for one CPU shows CPU efficiency versus domain size, the latter being proportional to the vector length. For the turbulence applications under consideration in this report, the relevant application regime starts at grid sizes larger than 10^6 points. As expected, the performance increases with increasing vector length and saturates at an efficiency

Table 2. Maximum theoretical performance in MLup/s if limited by peak performance (Peak) or memory bandwidth (MemBW). The last two column presents the measured (domain size of 128^3)

Platform	max. MLup/s Peak	MemBW	BEST MLUPS
Intel Itanium 2 (1.6 GHz)	32.0	14.0	8.89
NEC SX-6+ (565 MHz)	45.0	118	37.5
NEC SX-8 (2 GHz)	80.0	210	66.8

of close to 75%, i.e. at a single processor application performance of 11.9 GFlop/s. Note that this is equivalent to 68 MLup/s.

For a parallel scalability analysis we focus on weak-scaling scenarios which are typical for turbulence applications, where the total domain size should be as large as possible. In this case the total grid size per processor is kept constant which means that the overall problem size increases linearly with the number of CPUs used for the run. Furthermore, also the ratio of communication and computation remains constant. The inter-node parallelization implements domain decomposition and MPI for message passing. Computation and communication are completely separated by introducing additional communication layers around the computational domain. Data exchange between halo cells is done by `mpi_sendrecv` with local copying.

In Fig. 3 we show weak scaling results for the NEC SX-8 with up to 576 CPUs. In the chosen notation, perfect linear speedup would be indicated by all curves collapsing with the single CPU measurements.

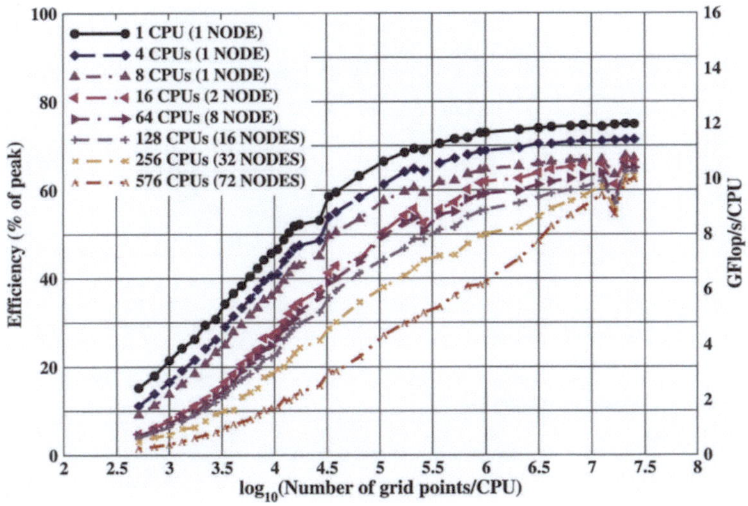

Fig. 3. Efficiency, GFlop/s and MLup/s of the lattice Boltzmann solver *BEST* depending on the domain size and the number of processors for up to 72 nodes NEC SX-8

On the SX-8, linear speedup can be observed for up to 4 CPUs. For 8 CPUs we find a performance degradation of about 10% per CPU which is, however, still far better than the numbers that can be achieved on bus based SMP systems, e.g. Intel Xeon or Intel Itanium 2. For intra-node communication the effect gets worse at intermediate problem sizes. In the case of inter-node jobs which use more than 8 CPUs it should be mentioned that the message sizes resulting from the chosen domain decomposition are too short to yield maximum bandwidth.

Figure 4 shows efficiency and performance numbers for an SGI Altix3700Bx2. On the Itanium 2 a slightly modified version of the collision-propagation loop is used which enables the compiler to software pipeline the loop. This implementation requires a substantially larger number of floating point operations per lattice site update than the vector implementation but performs best on cache based architectures, even if the number of lattice site updates per second is the measure. For more details we refer to Wellein et al. [21, 12]. The Itanium 2 achieves its maximum at 36% efficiency corresponding 2.3 GFlop/s or 9.3 MLup/s. Performance drops significantly when the problem size exceeds the cache size. Further increasing the problem size, compiler-generated prefetching starts to kick in and leads to gradual improvement up to a final level of 2.2 GFlop/s or 8.8 MLup/s. Unfortunately, when using both processors of a node, single CPU performance drops to 5.2 MLup/s. Going beyond two processors however, the NUMALink network is capable of almost perfectly scaling the single node performance over a wide range of problem sizes.

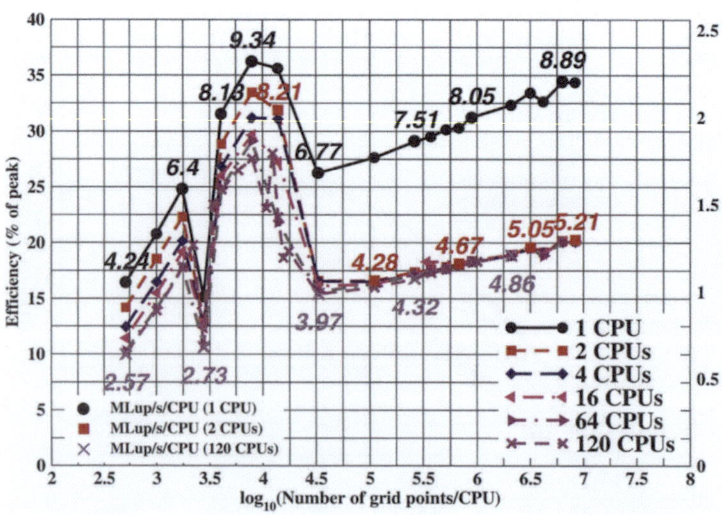

Fig. 4. Left: Efficiency, GFlop/s and MLup/s of the lattice Boltzmann solver $\mathcal{BEST}$ depending on the domain size and the number of processors for up to 120 CPUs an SGI Altix3700Bx2

5 Summary

Using a finite-volume and a lattice Boltzmann method (LBM) application we have demonstrated that the latest NEC SX-8 vector computer generation provides unmatched performance levels for applications which are data and computationally intensive. Another striking feature of the NEC vector series has also been clearly demonstrated: Going from the predecessor vector technology (SX-6+) to the SX-8 we found a performance improvement of roughly 1.7 which is the same as the ratio of the peak performance numbers (see Table 3).

To comment on the long standing discussion about the success of cache based microprocessors we have compared the NEC results with the SGI Altix system, being one of the best performing microprocessor systems for the applications under review here. We find that the per processor performance is in average almost one order of magnitude higher for the vector machine, clearly demonstrating that the vectors still provide a class of their own if application performance for vectorizable problems is the measure.

The extremely good single processor performance does not force the scientist to scale their codes and problems to thousands of processors in order to reach the Teraflop regime: For the LBM application we run a turbulence problem on a 576 processor NEC SX-8 system with a sustained performance of 5.7 TFlop/s. The same performance level would be require at least 6400 Itanium 2 CPUs on an SGI Altix3700.

Finally it should be emphasized that there has been a continuity of the basic principles of vector processor architectures for more than 20 years. This has provided highly optimized applications and solid experience in vector processor code tuning. Thus, the effort to benefit from technology advancements is minimal from a user's perspective. For the microprocessors, on the other hand, we suffer from a lack of continuity even on much smaller timescales. In the past years we have seen the rise of a completely new architecture (Intel Itanium). With the introduction of dual-/multi-core processors a new substantial change is just ahead, raising the question whether existing applications and conventional programming approaches are able to transfer the technological advancements of the "killer micros" to application performance.

Table 3. Typical performance ratios for the applications and computer architectures under consideration in this report. For the $\mathcal{LESOCC}$ we use the GFlop/s ratios and for $\mathcal{BEST}$ our results are based on MLups. Bandwidth restrictions due to the design of the SMP nodes have been incorporated as well

Performance Ratio	Peak Performance	$\mathcal{LESOCC}$	$\mathcal{BEST}$
NEC SX-8 vs. NEC SX-6+	1.77	1.71	1.78
NEC SX-8 vs. SGI Altix (1.6 GHz)	2.50	7.5–9	12–13

Acknowledgements

This work was financially supported by the High performance computer competence center Baden-Wuerttemberg[4] and by the Competence Network for Technical and Scientific High Performance Computing in Bavaria KONWIHR[5].

References

1. Brooks, E.: The attack of the killer micros. Teraflop Computing Panel, Supercomputing '89 (1989) Reno, Nevada, 1989
2. Oliker, L., A. Canning, J.C., Shalf, J., Skinner, D., Ethier, S., Biswas, R., Djomehri, J., d. Wijngaart, R.V.: Evaluation of cache-based superscalar and cacheless vector architectures for scientific computations. In: Proceedings of SC2003. CD-ROM (2003)
3. Oliker, L., A. Canning, J.C., Shalf, J., Ethier, S.: Scientific computations on modern parallel vector systems. In: Proceedings of SC2004. CD-ROM (2004)
4. Pohl, T., Deserno, F., Thürey, N., Rüde, U., Lammers, P., Wellein, G., Zeiser, T.: Performance evaluation of parallel large-scale lattice Boltzmann applications on three supercomputing architectures. In: Proceedings of SC2004. CD-ROM (2004)
5. HLRS/NEC: Teraflop workbench. http://www.teraflop-workbench.de/ (2005)
6. Breuer, M., Rodi, W.: Large–eddy simulation of complex turbulent flows of practical interest. In Hirschel, E.H., ed.: Flow Simulation with High–Performance Computers II. Volume 52., Vieweg Verlag, Braunschweig (1996) 258–274
7. Breuer, M.: Large–eddy simulation of the sub–critical flow past a circular cylinder: Numerical and modeling aspects. Int. J. for Numer. Methods in Fluids **28** (1998) 1281–1302
8. Breuer, M.: A challenging test case for large–eddy simulation: High Reynolds number circular cylinder flow. Int. J. of Heat and Fluid Flow **21** (2000) 648–654
9. Breuer, M.: Direkte Numerische Simulation und Large–Eddy Simulation turbulenter Strömungen auf Hochleistungsrechnern. Berichte aus der Strömungstechnik, Habilitationsschrift, Universität Erlangen–Nürnberg, Shaker Verlag, Aachen (2002) ISBN: 3–8265–9958–6.
10. Lammers, P.: Direkte numerische Simulationen wandgebundener Strömungen kleiner Reynoldszahlen mit dem lattice Boltzmann Verfahren. Dissertation, Universität Erlangen–Nürnberg (2005)
11. Deserno, F., Hager, G., Brechtefeld, F., Wellein, G.: Performance of scientific applications on modern supercomputers. In Wagner, S., Hanke, W., Bode, A., Durst, F., eds.: High Performance Computing in Science and Engineering, Munich 2004. Transactions of the Second Joint HLRB and KONWIHR Result and Reviewing Workshop, March 2nd and 3rd, 2004, Technical University of Munich. Springer Verlag (2004) 3–25
12. Wellein, G., Zeiser, T., Donath, S., Hager, G.: On the single processor performance of simple lattice Boltzmann kernels. Computers & Fluids (in press, available online December 2005)
13. Smagorinsky, J.: General circulation experiments with the primitive equations, I, the basic experiment. Mon. Weather Rev. **91** (1963) 99–165

[4] `http://www.hkz-bw.de/`

[5] `http://konwihr.in.tum.de/index_e.html`

14. Germano, M., Piomelli, U., Moin, P., Cabot, W.H.: A dynamic subgrid scale eddy viscosity model. Phys. of Fluids A **3** (1991) 1760–1765
15. Lilly, D.K.: A proposed modification of the Germano subgrid scale closure method. Phys. of Fluids A **4** (1992) 633–635
16. Stone, H.L.: Iterative solution of implicit approximations of multidimensional partial differential equations. SIAM J. Num. Anal. **91** (1968) 530–558
17. Deserno, F., Hager, G., Brechtefeld, F., Wellein, G.: Basic Optimization Strategies for CFD-Codes. Technical report, Regionales Rechenzentrum Erlangen (2002)
18. Qian, Y.H., d'Humières, D., Lallemand, P.: Lattice BGK models for Navier-Stokes equation. Europhys. Lett. **17** (1992) 479–484
19. Wolf-Gladrow, D.A.: Lattice-Gas Cellular Automata and Lattice Boltzmann Models. Volume 1725 of Lecture Notes in Mathematics. Springer, Berlin (2000)
20. Succi, S.: The Lattice Boltzmann Equation – For Fluid Dynamics and Beyond. Clarendon Press (2001)
21. Wellein, G., Lammers, P., Hager, G., Donath, S., Zeiser, T.: Towards optimal performance for lattice boltzmann applications on terascale computers. In: Parallel Computational Fluid Dynamics 2005, Trends and Applications. Proceedings of the Parallel CFD 2005 Conference, May 24–27, Washington D. C., USA. (2005) submitted.

Performance and Applications on Vector Systems

Performance Evaluation of Lattice-Boltzmann Magnetohydrodynamics Simulations on Modern Parallel Vector Systems

Jonathan Carter and Leonid Oliker

NERSC/CRD, Lawrence Berkeley National Laboratory, Berkeley, CA 94720, USA,
{jtcarter,loliker}@lbl.gov

Abstract The last decade has witnessed a rapid proliferation of superscalar cache-based microprocessors to build high-end computing (HEC) platforms, primarily because of their generality, scalability, and cost effectiveness. However, the growing gap between sustained and peak performance for full-scale scientific applications on such platforms has become major concern in high performance computing. The latest generation of custom-built parallel vector systems have the potential to address this concern for numerical algorithms with sufficient regularity in their computational structure. In this work, we explore two and three dimensional implementations of a lattice-Boltzmann magnetohydrodynamics (MHD) physics application, on some of today's most powerful supercomputing platforms. Results compare performance between the the vector-based Cray X1, Earth Simulator, and newly-released NEC SX-8, with the commodity-based superscalar platforms of the IBM Power3, Intel Itanium2, and AMD Opteron. Overall results show that the SX-8 attains unprecedented aggregate performance across our evaluated applications.

1 Introduction

The last decade has witnessed a rapid proliferation of superscalar cache-based microprocessors to build high-end computing (HEC) platforms. This is primarily because their generality, scalability, and cost effectiveness convinced computer vendors and users that vector architectures hold little promise for future large-scale supercomputing systems. However, the constant degradation of superscalar sustained performance has become a well-known problem in the scientific computing community. This trend has been widely attributed to the use of superscalar-based commodity components whose architectural designs offer a balance between memory performance, network capability, and execution rate, that is poorly matched to the requirements of large-scale numerical computations. The latest generation of custom-built parallel vector systems are addressing these challenges for numerical algorithms amenable to vectorization.

Superscalar architectures are unable to efficiently exploit the large number of floating-point units that can be potentially fabricated on a chip, due to the small granularity of their instructions and the correspondingly complex control structure necessary to support it. Vector technology, on the other hand, provides an efficient approach for controlling a large amount of computational resources provided that sufficient regularity in the computational structure can be discovered. Vectors exploit these regularities to expedite uniform operations on independent data elements, allowing memory latencies to be masked by overlapping pipelined vector operations with memory fetches. Vector instructions specify a large number of identical operations that may execute in parallel, thus reducing control complexity and efficiently controlling a large amount of computational resources. However, when such operational parallelism cannot be found, the efficiency of the vector architecture can suffer from the properties of Amdahl's Law, where the time taken by the portions of the code that are non-vectorizable easily dominate the execution time.

In order to quantify what modern vector capabilities entail for the scientific communities that rely on modeling and simulation, it is critical to evaluate them in the context of demanding computational algorithms. This work compares performance between the vector-based Cray X1, Earth Simulator (ES) and newly-released NEC SX-8, with commodity-based superscalar platforms: the IBM Power3, Intel Itanium2, and AMD Opteron. We study the behavior of two scientific codes with the potential to run at ultra-scale, in the areas of magnetohydrodynamics (MHD) physics simulations (LBMHD2D and LBMHD3D). Our work builds on our previous efforts [1, 2] and makes the contribution of adding recently acquired performance data for the SX-8, and the latest generation of superscalar processors. Additionally, we explore improved vectorization techniques for LBMHD2D. Overall results show that the SX-8 attains unprecedented aggregate performance across our evaluated applications, continuing the trend set by the ES in our previous performance studies.

2 HEC Platforms and Evaluated Applications

In this section we briefly describe the computing platforms and scientific applications examined in our study. Tables 1 and 2 present an overview of the salient features for the six parallel HEC architectures. Observe that the vector machines

Table 1. CPU overview of the Power3, Itanium2, Opteron, X1, ES, and SX-8 platforms

Platform	CPU/ Node	Clock (MHz)	Peak (GF/s)	Mem BW (GB/s)	Peak (Byte/Flop)
Power3	16	375	1.5	0.7	0.47
Itanium2	4	1400	5.6	6.4	1.1
Opteron	2	2200	4.4	6.4	1.5
X1	4	800	12.8	34.1	2.7
ES (Modified SX-6)	8	500	8.0	32.0	4.0
SX-8	8	2000	16.0	64.0	4.0

Table 2. Interconnect performance of the Power3, Itanium2, Opteron, X1, ES, and SX-8 platforms

Platform	Network	MPI Lat (μsec)	MPI BW (GB/s/CPU)	Bisect BW (Byte/Flop)	Network Topology
Power3	Colony	16.3	0.13	0.09	Fat-tree
Itanium2	Quadrics	3.0	0.25	0.04	Fat-tree
Opteron	InfiniBand	6.0	0.59	0.11	Fat-tree
X1	Custom	7.3	6.3	0.09	2D-torus
ES (Modified SX-6)	Custom (IN)	5.6	1.5	0.19	Crossbar
SX-8	IXS	5.0	2.0	0.13	Crossbar

have higher peak performance and better system balance than the superscalar platforms. Additionally, the X1, ES, and SX-8 have high memory bandwidth relative to peak CPU speed (bytes/flop), allowing them to more effectively feed the arithmetic units. Finally, the vector platforms utilize interconnects that are tightly integrated to the processing units, with high performance network buses and low communication software overhead.

Three superscalar commodity-based platforms are examined in our study. The IBM Power3 experiments reported were conducted on the 380-node IBM pSeries system, Seaborg, running AIX 5.2 (Xlf compiler 8.1.1) and located at Lawrence Berkeley National Laboratory (LBNL). Each SMP node consists of sixteen 375 MHz processors (1.5 Gflop/s peak) connected to main memory via the Colony switch using an omega-type topology. The AMD Opteron system, Jacquard, is also located at LBNL and contains 320 dual nodes, running Linux 2.6.5 (PathScale 2.0 compiler). Each node contains two 2.2 GHz Opteron processors (4.4 Gflop/s peak), interconnected via Infiniband fabric in a fat-tree configuration. Finally, the Intel Itanium experiments were performed on the Thunder system, consisting of 1024 nodes, each containing four 1.4 GHz Itanium2 processors (5.6 Gflop/s peak) and running Linux Chaos 2.0 (Fortran version ifort 8.1). The system is interconnected using Quadrics Elan4 in a fat-tree configuration, and is located at Lawrence Livermore National Laboratory.

We also examine three state-of-the-art parallel vector systems. The Cray X1 is designed to combine traditional vector strengths with the generality and scalability features of modern superscalar cache-based parallel systems. The computational core, called the single-streaming processor (SSP), contains two 32-stage vector pipes running at 800 MHz. Each SSP contains 32 vector registers holding 64 double-precision words, and operates at 3.2 Gflop/s peak for 64-bit data. The SSP also contains a two-way out-of-order superscalar processor running at 400 MHz with two 16KB caches (instruction and data). Four SSP can be combined into a logical computational unit called the multi-streaming processor (MSP) with a peak of 12.8 Gflop/s. The four SSPs share a 2-way set associative 2MB data Ecache, a unique feature for vector architectures that allows extremely high bandwidth (25–51 GB/s) for computations with temporal data locality. The X1 node consists of four MSPs sharing a flat memory, and large system configuration are networked through a modified 2D torus interconnect. All reported X1 experiments were performed on the 512-MSP system (several reserved for system services) running UNICOS/mp 2.5.33 (5.3 programming environment) and operated by Oak Ridge National Laboratory.

The vector processor of the ES uses a dramatically different architectural approach than conventional cache-based systems. Vectorization exploits regularities in the computational structure of scientific applications to expedite uniform operations on independent data sets. The 500 MHz ES processor is an enhanced NEC SX6, containing an 8-way replicated vector pipe with a peak performance of 8.0 Gflop/s per CPU. The Earth Simulator is the world's third most powerful supercomputer [3], contains 640 ES nodes connected through a custom single-stage IN crossbar. The 5120-processor ES runs Super-UX, a 64-bit Unix operating system based on System V-R3 with BSD4.2 communication features. As remote ES access is not available, the reported experiments were performed during the authors' visit to the Earth Simulator Center located in Kanazawa-ku, Yokohama, Japan in 2003 and 2004.

Finally, we examine the newly-released NEC SX-8, currently the world's most powerful vector processor. The SX-8 architecture operates at 2 GHz, and contains four replicated vector pipes for a peak performance of 16 Gflop/s per processor. The SX-8 architecture has several enhancements compared with the ES/SX6 predecessor, including improved divide performance, hardware square root functionality, and in-memory caching for reducing bank conflict overheads. However, the SX-8 used in our study uses commodity DDR-SDRAM; thus, we expect higher memory overhead for irregular accesses when compared with the specialized high-speed FPLRAM (Full Pipelined RAM) of the ES. Both the ES and SX-8 processors contain 72 vector registers each holding 256 doubles, and utilize scalar units operating at the half the peak of their vector counterparts. All reported SX-8 results were run on the 36 node (72 are currently to be available) system located at High Performance Computer Center (HLRS) in Stuttgart, Germany. This HLRS SX-8 is interconnected with the NEC Custom IXS network and runs Super-UX (Fortran Version 2.0 Rev.313).

3 Magnetohydrodynamic Turbulence Simulation

Lattice Boltzmann methods (LBM) have proved a good alternative to conventional numerical approaches for simulating fluid flows and modeling physics in fluids [4]. The basic idea of the LBM is to develop a simplified kinetic model that incorporates the essential physics, and reproduces correct macroscopic averaged properties. Recently, several groups have applied the LBM to the problem of magnetohydrodynamics (MHD) [5, 6] with promising results. We use two LB MHD codes, a previously used 2D code [7, 1] and a more recently developed 3D code. In both cases, the codes simulate the behavior of a conducting fluid evolving from simple initial conditions through the onset of turbulence. Figure 1 shows a slice through the xy-plane in the (left) 2D and right (3D) simulation, where the vorticity profile has considerably distorted after several hundred time

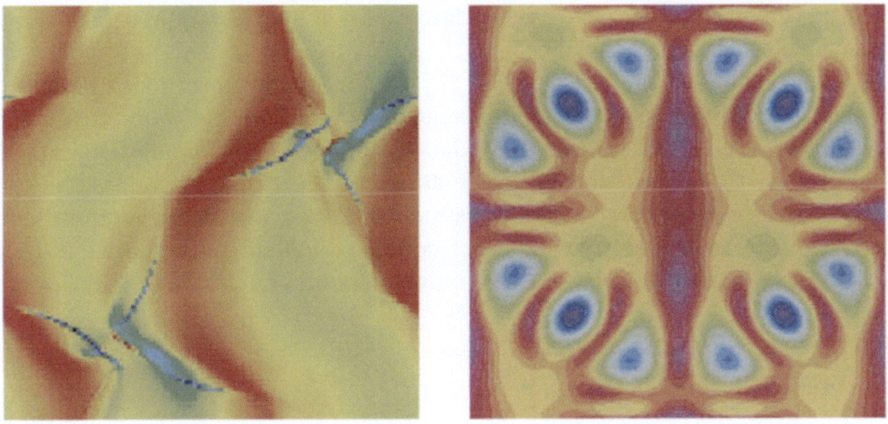

Fig. 1. Contour plot of xy-plane showing the evolution of vorticity from well-defined tube-like structures into turbulent structures using (left) LBMHD2D and (right) LBMHD3D

steps as computed by LBMHD. In the 2D case, the square spatial grid is coupled to an octagonal streaming lattice and block distributed over a 2D processor grid as shown in Fig. 2. The 3D spatial grid is coupled via a 3DQ27 streaming lattice and block distributed over a 3D Cartesian processor grid. Each grid point is associated with a set of mesoscopic variables, whose values are stored in vectors proportional to the number of streaming directions – in this case 9 and 27 (8 and 26 plus the null vector).

The simulation proceeds by a sequence of collision and stream steps. A collision step involves data local only to that spatial point, allowing concurrent, dependence-free point updates; the mesoscopic variables at each point are updated through a complex algebraic expression originally derived from appropriate conservation laws. A stream step evolves the mesoscopic variables along the streaming lattice, necessitating communication between processors for grid points at the boundaries of the blocks.

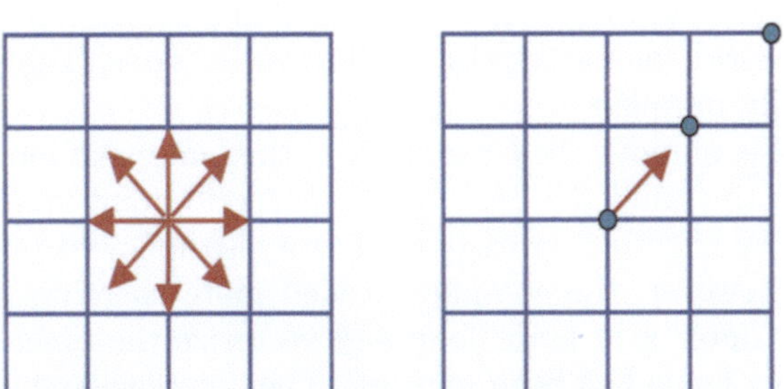

Fig. 2. Octagonal streaming lattice superimposed over a square spatial grid (left) requires diagonal velocities to be interpolated onto three spatial gridpoints (right)

Additionally, for the 2D case, an interpolation step is required between the spatial and streaming lattices since they do not match. This interpolation is folded into the stream step. For the 3D case, a key optimization described by Wellein and co-workers [8] was implemented, saving on the work required by the stream step. They noticed that the two phases of the simulation could be combined, so that either the newly calculated particle distribution function could be scattered to the correct neighbor as soon as it was calculated, or equivalently, data could be gathered from adjacent cells to calculate the updated value for the current cell. Using this strategy, only the points on cell boundaries require copying.

3.1 Vectorization Details

The basic computational structure consists of two or three nested loops over spatial grid points (typically 1000s iterations) with inner loops over velocity streaming vectors and magnetic field streaming vectors (typically 10–30 iterations), performing various algebraic expressions. Although the two codes have kernels which are quite similar, our experiences in optimizing were somewhat different.

For the 2D case, in our earlier work on the ES, attempts to make the compiler vectorize the inner gridpoint loops rather than the streaming loops failed. The inner grid point loop was manually taken inside the streaming loops, which were hand unrolled twice in the case of small loop bodies. In addition, the array temporaries added were padded to reduce bank conflicts. With the hindsight of our later 3D code experience, this strategy is clearly not optimal. Better utilization of the multiple vector pipes can be achieved by completely unrolling the streaming loops and thus increasing the volume of work within the vectorized loops. We have verified that this strategy does indeed give better performance than the original algorithm on both the ES and SX-8, and show results that illustrate this in the next section. Turning to the X1, the compiler did an excellent job, multi-streaming the outer grid point loop and vectorizing the inner grid point loop after unrolling the stream loops without any user code restructuring. For the superscalar architectures some effort was made to tune for better cache use. First, the inner gridpoint loop was blocked and inserted into the streaming loops to provide stride-one access in the innermost loops. The streaming loops were then partially unrolled.

For the 3D case, on both the ES and SX-8, the innermost loops were unrolled via compiler directives and the (now) innermost grid point loop was vectorized. This proved a very effective strategy, and was also followed on the X1. In the case of the X1, however, the compiler needed more coercing via directives to multi-stream the outer grid point loop and vectorize the inner grid point loop once the streaming loops had been unrolled. The difference in behavior is clearly related to the size of the unrolled loop body, the 3D case being a factor of approximately three more complicated. In the case of X1 the number of vector registers available for a vectorized loop is more limited than for the SX systems and for complex loop bodies register spilling will occur. However, in this case, the strategy pays off as shown experimental results section below. For the superscalar architectures, we utilized a data layout that has been previously shown to be optimal on cache-based machines [8], but did not explicitly tune for the cache size on any machine.

Interprocessor communication was implemented using the MPI library, by copying the non-contiguous mesoscopic variables data into temporary buffers, thereby reducing the required number of send/receive messages.

These codes represent candidate ultra-scale applications that have the potential to fully utilize leadership-class computing systems. Performance results, presented in Gflop/s per processor and percentage of peak, are used to compare the relative time to solution of our evaluated computing systems. Since different algorithmic approaches are used for the vector and scalar implementations, this value is computed by dividing a baseline flop-count (as measured on the ES) by the measured wall-clock time of each platform.

3.2 Experimental Results

Tables 3 and 4 present the performance of both LBMHD applications across the six architectures evaluated in our study. Cases where the memory required exceeded that available are indicated with a dash. For LBMHD2D we show the performance of both vector algorithms (first strip-mined as used in the original ES experiment, and second using the new unrolled inner loop) for the SX-8. In accordance with the discussion in the previous section, the new algorithm clearly outperforms the old.

Table 3. LBMHD2D performance in GFlop/s (per processor) across the studied architectures for a range of concurrencies and grid sizes. The original and optimized algorithms are shown for the ES and SX-8. Percentage of peak is shown in parenthesis

P	Size	Power3	Itanium2	Opteron	X1	original ES	optimized ES	original SX-8	optimized SX-8
16	4096^2	0.11 (7)	0.40 (7)	0.83 (19)	4.32 (34)	4.62 (58)	5.00 (63)	6.33 (40)	7.45 (47)
64	4096^2	0.14 (9)	0.42 (7)	0.81 (18)	4.35 (34)	4.29 (54)	4.36 (55)	4.75 (30)	6.28 (39)
64	8192^2	0.11 (7)	0.40 (7)	0.81 (18)	4.48 (35)	4.64 (58)	5.01 (62)	6.01 (38)	7.03 (44)
256	8192^2	0.12 (8)	0.38 (6)		2.70 (21)	4.26 (53)	4.43 (55)	4.44 (28)	5.51 (34)

Table 4. LBMHD3D performance in GFlop/s (per processor) across the studied architectures for a range of concurrencies and grid sizes. Percentage of peak is shown in parenthesis

P	Size	Power3	Itanium2	Opteron	X1	ES	SX-8
16	256^3	0.14 (9)	0.26 (5)	0.70 (16)	5.19 (41)	5.50 (69)	7.89 (49)
64	256^3	0.15 (10)	0.35 (6)	0.68 (15)	5.24 (41)	5.25 (66)	8.10 (51)
256	512^3	0.14 (9)	0.32 (6)	0.60 (14)	5.26 (41)	5.45 (68)	9.66 (60)
512	512^3	0.14 (9)	0.35 (6)	0.59 (13)	–	5.21 (65)	–

Observe that the vector architectures clearly outperform the scalar systems by a significant factor. Across these architectures, the LB applications exhibit an average vector length (AVL) very close to the maximum and a very high vector operation ratio (VOR). In absolute terms, the SX-8 is the leader by a wide margin, achieving the highest per processor performance to date for LBMHD3D. The ES, however, sustains the highest fraction of peak across all architectures – 65% even at the highest 512-processor concurrency. Examining the X1 behavior, we see that in MSP mode absolute performance is similar to the ES. The high performance of the X1 is gratifying since we noted several outputed warnings concerning vector register spilling during the optimization of the collision routine. Because the X1 has fewer vector registers than the ES/SX-8 (32 vs 72), vectorizing these complex loops will exhaust the hardware limits and force spilling to memory. That we see no performance penalty is probably due to the spilled registers being effectively cached.

Turning to the superscalar architectures, the Opteron cluster outperforms the Itanium2 system by almost a factor of 2X. One source of this disparity is that the 2-way SMP Opteron node has STREAM memory bandwidth [9] of more than twice that of the Itanium2 [10], which utilizes a 4-way SMP node configuration. Another possible source of this degradation are the relatively high cost of inner-loop register spills on the Itanium2, since the floating point values cannot be stored in the first level of cache. Given the age and specifications, the Power3 does quite reasonably, obtaining a higher percent of peak that the Itanium2, but falling behind the Opteron.

Although the SX-8 achieves the highest absolute performance, the percentage of peak is somewhat lower than that of ES. We believe that this is related to the memory subsystem and use of commodity DDR-SDRAM. In order to test this hypothesis, we recorded the time due to memory bank conflicts for both applications on the ES and SX-8 using the ftrace tool, and present it in Table 5.

Most obviously in the case of the 2D code, the amount of time spent due to bank conflicts is appreciably larger for the SX-8. Efforts to reduce the amount of time for bank conflicts for the 2D 64 processor benchmark produced a slight improvement to 13%. In the case of the 3D code, the effects of bank conflicts are minimal.

Table 5. LBMHD2D and LBMHD3D bank conflict time (as percentage of real time) shown for a range of concurrencies and grid sizes on ES and SX-8

Code	*P*	Grid Size	ES BC (%)	SX-8 BC (%)
2D	64	8192^2	0.3	16.6
2D	256	8192^2	0.3	10.7
3D	64	256^3	>0.01	1.1
3D	256	512^3	>0.01	1.2

4 Conclusions

This study examined two scientific codes on the parallel vector architectures of the X1, ES and SX-8, and three superscalar platforms, Power3, Itanium2, and Opteron. A summary of the results for the largest comparable problem size and concurrency is shown in Fig. 3, for both (left) raw performance and (right) percentages of peak. Overall results show that the SX-8 achieved the highest performance of any architecture tested to date, demonstrating the tremendous potential of modern parallel vector systems. However, the SX-8 could not match the sustained performance of the ES, due in part, to a relatively higher memory latency overhead for irregular data accesses. Both the SX-8 and ES also consistently achieved a significantly higher fraction of peak than the X1, due to superior scalar processor performance, memory bandwidth, and network bisection bandwidth relative to the peak vector flop rate. Finally, a comparison of the superscalar platforms shows that the Opteron consistently outperformed the Itanium2 and Power3, both in terms of raw speed and efficiency – due, in part, to its on-chip memory controller and (unlike the Itanium2) the ability to store floating point data in the L1 cache. The Itanium2 exceeds the performance of the (relatively old) Power3 processor, however its obtained percentage of peak falls further behind. Future work will expand our study to include additional areas of computational sciences, while examining the latest generation of supercomputing platforms, including BG/L, X1E, and XT3.

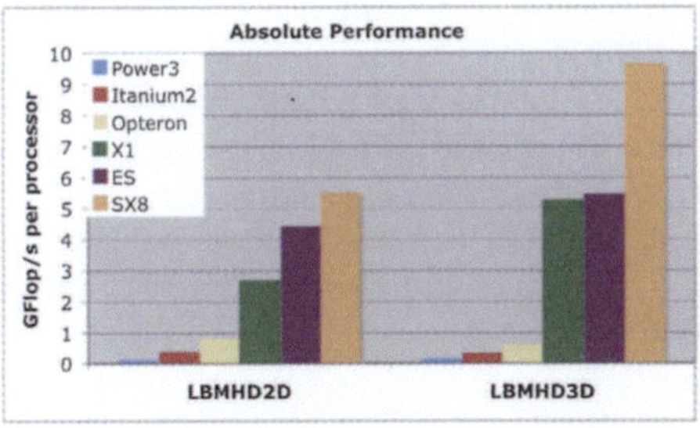

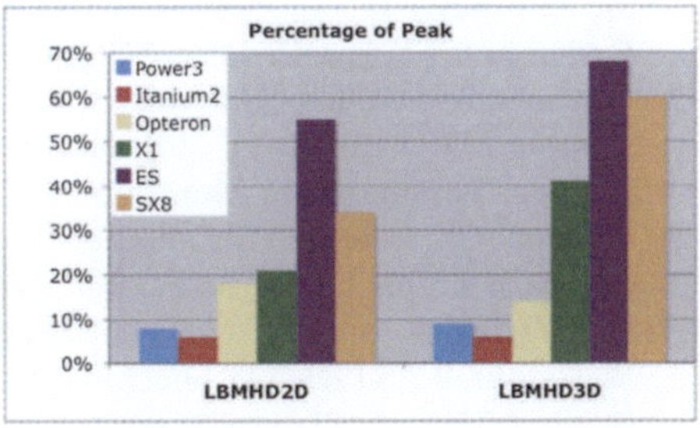

Fig. 3. Summary comparison of (left) raw performance and (right) percentage of peak across our set of evaluated applications and architectures

Acknowledgements

The authors would like to thank the staff of the Earth Simulator Center, especially Dr. T. Sato, S. Kitawaki and Y. Tsuda, for their assistance during our visit. We are also grateful for the early SX-8 system access provided by HLRS, Germany. This research used resources of the National Energy Research Scientific Computing Center, which is supported by the Office of Science of the U.S. Department of Energy under Contract No. DE-AC02-05CH11231. This research used resources of the Lawrence Livermore National Laboratory, which is supported by the Office of Science of the U.S. Department of Energy under contract No. W-7405-Eng-48. This research used resources of the Center for Computational Sciences at Oak Ridge National Laboratory, which is supported by the Office of Science of the Department of Energy under Contract DE-AC05-00OR22725. LBNL authors were supported by the Office of Advanced Scientific Computing Research in the Department of Energy Office of Science under contract number DE-AC02-05CH11231.

References

1. Oliker, L., Canning, A., Carter, J., Shalf, J., Ethier, S.: Scientific computations on modern parallel vector systems. In: Proc. SC2004: High performance computing, networking, and storage conference. (2004)
2. Oliker, L. et al.: Evaluation of cache-based superscalar and cacheless vector architectures for scientific computations. In: Proc. SC2003: High performance computing, networking, and storage conference. (2003)
3. Meuer, H., Strohmaier, E., Dongarra, J., Simon, H.: Top500 Supercomputer Sites. (`http://www.top500.org`)
4. Succi, S.: The lattice Boltzmann equation for fluids and beyond. Oxford Science Publ. (2001)
5. Dellar, P.: Lattice kinetic schemes for magnetohydrodynamics. J. Comput. Phys. **79** (2002)
6. Macnab, A., Vahala, G., Pavlo, P., Vahala, L., Soe, M.: Lattice Boltzmann model for dissipative incompressible MHD. In: Proc. 28th EPS Conference on Controlled Fusion and Plasma Physics. Volume 25A. (2001)
7. Macnab, A., Vahala, G., Vahala, L., Pavlo, P.: Lattice boltzmann model for dissipative MHD. In: Proc. 29th EPS Conference on Controlled Fusion and Plasma Physics. Volume 26B., Montreux, Switzerland (June 17–21, 2002)
8. Wellein, G., Zeiser, T., Donath, S., Hager, G.: On the single processor performance of simple lattice bolzmann kernels. Computers and Fluids (Article in press, http://dx.doi.org/10.1016/j.compfluid.2005.02.008)
9. McCalpin, J.: STREAM benchmark. (`http://www.cs.virginia.edu/stream/ref.html`)
10. Dongarra, J., Luszczek, P.: HPC challenge benchmark. (`http://icl.cs.utk.edu/hpcc/index.html`)

Over 10 TFLOPS Computation for a Huge Sparse Eigensolver on the Earth Simulator

Toshiyuki Imamura[1], Susumu Yamada[2], and Masahiko Machida[2,3]

[1] Department of Computer Science, the University of Electro-Communications, 1-5-1 Chofugaoka, Chofu-shi, Tokyo 182-8585, Japan, `imamura@im.uec.ac.jp`,
[2] Center for Computational Science and Engineering, Japan Atomic Energy Agency, 6-9-3 Higashi-Ueno, Taitoh-ku, Tokyo 110-0015, Japan, {`yamada.susumu`, `machida.masahiko`}`@jaea.go.jp`,
[3] CREST, JST, 4-1-8, Honcho, Kawaguchi-shi, Saitama 330-0012, Japan

Abstract To investigate a possibility of special physical properties like superfluidity, we implement a high performance exact diagonalization code for the trapped Hubbard model on the Earth Simulator. From the numerical and computational point of view, it is found that the performance of the preconditioned conjugate gradient (PCG) method is excellent in our case. It is 1.5 times faster than the conventional Lanczos one since it can conceal the communication overhead much more effectively. Consequently, the PCG method shows 16.14 TFLOPS on 512 nodes. Furthermore, we succeed in solving a 120-billion-dimensional matrix. To our knowledge, this dimension is a world-record.

1 Introduction

The condensation in fermion system is one of the most universal issues in fundamental physics, since particles which form matters, i.e., electron, proton, neutron, quark, and so on, are all fermions. Motivated by such broad background, we numerically explore a possibility of superfluidity in the atomic Fermi gas [1]. Our undertaking model is the fermion-Hubbard model [2, 3] with trapping potential. The Hubbard model is one of the most intensively-studied models by computers because of its rich physics and quite simple model expression [2]. The Hamiltonian of the Hubbard model with a trap potential [1, 4] is given as (see details in the literature [1])

$$H_{\text{Hubbard}} = -t \sum_{i,j,\sigma} (a_{j,\sigma}^{\dagger} a_{i,\sigma} + \text{H.C.}) + U \sum_{i} n_{i\uparrow} n_{i\downarrow} + \left(\frac{2}{N}\right)^2 V \sum_{i,\sigma} n_{i,\sigma} \left(i - \frac{N}{2}\right)^2 . \quad (1)$$

The computational finite-size approaches on this model are roughly classified into two types, the exact diagonalization using the Lanczos method [5], and the quantum Monte Carlo [2]. The former directly calculates the ground and the low lying excited states of the model, and moreover, obtains various physical quantities with high accuracy. However, the numbers of fermions and sites are severely limited because the matrix size of the Hamiltonian grows exponentially with increasing these numbers. On the other hand, the latter has an advantage in terms of these numbers, but confronts a fatal problem because of the negative sign in the probability calculation [2]. In this study, we choose the conventional method, the exact diagonalization. One can raise a challenging theme for supercomputing, that is, to implement the exact diagonalization code on the present top-class supercomputer, i.e., the Earth Simulator [6], and to examine how large matrices can be solved and how excellent performance can be obtained.

In this paper, we develop a new type of high performance application which solves the eigenvalue problem of the Hubbard Hamiltonian matrix (1) on the Earth Simulator, and present the progress in numerical algorithm and software implementation to obtain the best performance exceeding 10 TFLOPS and solve the world-record class of large matrices.

The rest of this paper covers as follows. In Sect. 2, we briefly introduce the Earth Simulator, and two eigenvalue solvers to diagonalize the Hamiltonian matrix of the Hubbard model taking their convergence properties into consideration in Sect. 3. Section 4 presents the implementation of two solvers on the Earth Simulator, and Sect. 5 shows actual performance in large-scale matrix diagonalizations on the Earth Simulator.

2 The Earth Simulator

The Earth Simulator (hereafter ES), developed by NASDA (presently JAXA), JAERI (presently JAEA), and JAMSTEC, is situated on the flagship class of highly parallel vector supercomputer. The theoretical peak performance is 40.96 TFLOPS, and the total memory size is 10 TByte (see Table 1). The architecture of the ES is quite suitable for scientific and technological computation [6] due to well-balance of the processing speed of the floating point operation and the memory bandwidth as well as the network throughput. Therefore, several applications achieved excellent performance, and some of them won honorable awards. On the ES, one can naturally expect not only parametric surveys but also grand-challenge problems to innovate untouched scientific fields.

Our goals on this work are to support an advanced large-scale physical simulation, to achieve comparable high performance to the applications which won the Gordon Bell prize as shown in Table 1, and to illustrate better approaches of software implementation to obtain the best performance on the ES.

Table 1. Hardware configuration, and the best performed applications of the ES (at March, 2005)

The number of nodes	640 (8PE's/node, total 5120PE's)
PE	VU(Mul/Add)×8pipes, Superscalar unit
Main memory & bandwidth	10TB (16GB/node), 256GB/s/node
Interconnection	Metal-cable, Crossbar, 12.3GB/s/1way
Theoretical peak performance	40.96TFLOPS (64GFLOPS/node, 8GFLOPS/PE)
Linpack (TOP500 List)	35.86TFLOPS (87.5% of the peak) [7]
The fastest real application	26.58TFLOPS (64.9% of the peak) [8] Complex number calculation (mainly FFT)
Our goal	Over 10TFLOPS (32.0% of the peak) [9] Real number calculation (Numerical algebra)

3 Numerical Algorithms

The core of our program is to calculate the smallest eigenvalue and the corresponding eigenvector for $Hv = \lambda v$, where the matrix is real and symmetric. Several iterative numerical algorithms, i.e., the power method, the Lanczos method, the conjugate gradient method (CG), and so on, are available. Since the ES is public resource and a use of hundreds of nodes is limited, the most effective algorithm must be selected before large-scale simulations.

3.1 Lanczos Method

The Lanczos method is one of the subspace projection methods that creates a Krylov sequence and expands invariant subspace successively based on the procedure of the Lanczos principle [10] (see Fig. 1(a)). Eigenvalues of the projected invariant subspace well approximate those of the original matrix, and the subspace can be represented by a compact tridiagonal matrix. The main recurrence part of this algorithm repeats to generate the Lanczos vector v_{i+1} from v_{i-1} and v_i as seen in Fig. 1(a). In addition, an N-word buffer is required for storing an eigenvector. Therefore, the memory requirement is $3N$ words.

As shown in Fig 1(a), the number of iterations depends on the input matrix, however it is usually fixed by a constant number m. In the following, we choose a smaller empirical fixed number i.e., 200 or 300, as an iteration count.

3.2 Preconditioned Conjugate Gradient Method

Alternative projection method exploring invariant subspace, the conjugate gradient method is a popular algorithm, which is frequently used for solving linear systems. The algorithm is shown in Fig. 1(b), which is modified from the original algorithm [11] to reduce the load of the calculation S_A. This method has a lot of

(a)

$x_0 :=$ an initial guess.
$\beta_0 := 1, v_{-1} := 0, v_0 = x_0/\|x_0\|$
do i=0,1,..., $m-1$,
  or until $\beta_i < \epsilon$,
  $u_i := Hv_i - \beta_i v_{i-1}$
  $\alpha_i := (u_i, v_i)$
  $w_i := u_i - \alpha_i v_i$
  $\beta_i := \|w_i\|$
  $v_{i+1} := w_i/\beta_{i+1}$
enddo

(b)

$x_0 :=$ an initial guess., $p_0 := 0$,
$x_0 := x_0/\|x_0\|, X_0 := Hx_0, P_0 = 0$,
$\mu_{-1} := (x_0, X_0), w_0 := X_0 - \mu_{-1}x_0$
do i=0,1,..., until convergence
  $W_i := Hw_i$
  $S_A := \{w_i, x_i, p_i\}^T\{W_i, X_i, P_i\}$
  $S_B := \{w_i, x_i, p_i\}^T\{w_i, x_i, p_i\}$
  Solve the smallest eigenvalue μ
    and the corresponding vector v,
    $S_A v = \mu S_B v, v = (\alpha, \beta, \gamma)^T$.
  $\mu_i := (\mu + (x_i, X_i))/2$
  $x_{i+1} := \alpha w_i + \beta x_i + \gamma p_i, x_{i+1} := x_{i+1}/\|x_{i+1}\|$
  $p_{i+1} := \alpha w_i + \gamma p_i, p_{i+1} := p_{i+1}/\|p_{i+1}\|$
  $X_{i+1} := \alpha W_i + \beta X_i + \gamma P_i, X_{i+1} := X_{i+1}/\|x_{i+1}\|$
  $P_{i+1} := \alpha W_i + \gamma P_i, P_{i+1} := P_{i+1}/\|p_{i+1}\|$
  $w_{i+1} := T(X_{i+1} - \mu_i x_{i+1}), w_{i+1} := w_{i+1}/\|w_{i+1}\|$
enddo

Fig. 1. The Lanczos algorithm (left (**a**)), and the preconditioned conjugate gradient method (right (**b**))

advantages in the performance, because both the number of iterations and the total CPU time drastically decrease depending on the preconditioning [11]. The algorithm requires memory space to store six vectors, i.e., the residual vector w_i, the search direction vector p_i, and the eigenvector x_i, moreover, W_i, P_i, and X_i. Thus, the memory usage is totally $6N$ words.

In the algorithm illustrated in Fig. 1(b), an operator T indicates the preconditioner. The preconditioning improves convergence of the CG method, and its strength depends on mathematical characteristics of the matrix generally. However, it is hard to identify them in our case, because many unknown factor lies in the Hamiltonian matrix. Here, we focus on the following two simple preconditioners: point Jacobi, and zero-shift point Jacobi. The point Jacobi is the most classical preconditioner, and it only operates the diagonal scaling of the matrix. The zero-shift point Jacobi is a diagonal scaling preconditioner shifted by 'μ_k' to amplify the eigenvector corresponding to the smallest eigenvalue, i.e., the pre-

Table 2. Comparison among three preconditioners, and their convergence properties

	1) NP	2) PJ	3) ZS-PJ
Num. of Iterations	268	133	91
Residual Error	1.445E-9	1.404E-9	1.255E-9
Elapsed Time [sec]	78.904	40.785	28.205
FLOPS	382.55G	383.96G	391.37G

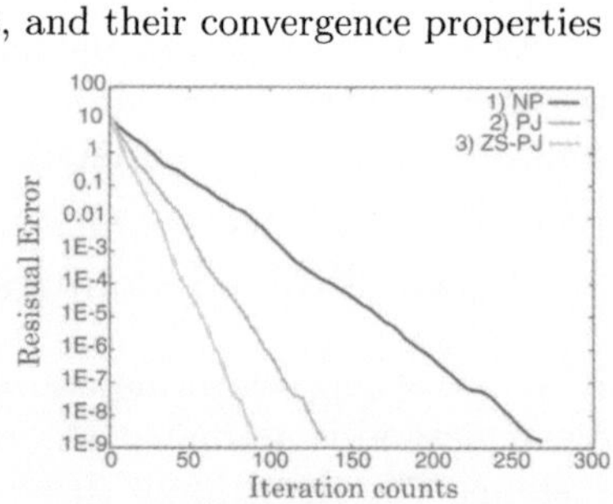

conditioning matrix is given by $T = (D - \mu_k I)^{-1}$, where μ_k is the approximate smallest eigenvalue which appears in the PCG iterations.

Table 2 summarizes a performance test of three cases, 1) without preconditioner (NP), 2) point Jacobi (PJ), and 3) zero-shift point Jacobi (ZS-PJ) on the ES, and the corresponding graph illustrates their convergence properties. Test configuration is as follows; 1,502,337,600-dimensional Hamiltonian matrix (12 fermions on 20 sites) and we use 10 nodes of the ES. These results clearly reveal that the zero-shift point Jacobi is the best preconditioner in this study.

4 Implementation on the Earth Simulator

The ES is basically classified in a cluster of SMP's which are interconnected by a high speed network switch, and each node comprises eight vector PE's. In order to achieve high performance in such an architecture, the intra-node parallelism, i.e., thread parallelization and vectorization, is crucial as well as the inter-node parallelization. In the intra-node parallel programming, we adopt the automatic parallelization of the compiler system using a special language extension. In the inter-node parallelization, we utilize the MPI library tuned for the ES. In this section, we focus on a core operation Hv common for both the Lanczos and the PCG algorithms and present the parallelization including data partitioning, the communication, and the overlap strategy.

4.1 Core Operation: Matrix-Vector Multiplication

The Hubbard Hamiltonian H (1) is mathematically given as

$$H = I \otimes A + A \otimes I + D, \tag{2}$$

where I, A, and D are the identity matrix, the sparse symmetric matrix due to the hopping between neighboring sites, and the diagonal matrix originated from the presence of the on-site repulsion, respectively.

Since the core operation Hv can be interpreted as a combination of the alternating direction operations like the ADI method which appears in solving a partial differential equation. In other word, it is transformed into the matrix-matrix multiplications as $Hv \mapsto (Dv, (I \otimes A)v, (A \otimes I)v) \mapsto (\bar{D} \odot V, AV, VA^T)$, where the matrix V is derived from the vector v by a two-dimensional ordering. The k-th element of the matrix D, d_k, is also mapped onto the matrix $\bar{D}$ in the same manner, and the operator $\odot$ means an element-wise product.

4.2 Data Distribution, Parallel Calculation, and Communication

The matrix A, which represents the site hopping of up (or down) spin fermions, is a sparse matrix. In contrast, the matrices V and $\bar{D}$ must be treated as dense matrices. Therefore, while all the CRS (Compressed Row Storage) format of the matrix A are stored on all the nodes, the matrices V and $\bar{D}$ are column-wisely partitioned among all the computational nodes. Moreover, the row-wisely partitioned V is also required on each node for parallel computing of VA^T. This means data re-distribution of the matrix V to V^T, that is the matrix transpose, and they also should be restored in the original distribution.

The core operation Hv including the data communication can be written as follows:

CAL1: $E^{\mathrm{col}} := \bar{D}^{\mathrm{col}} \odot V^{\mathrm{col}}$,
CAL2: $W_1^{\mathrm{col}} := E^{\mathrm{col}} + AV^{\mathrm{col}}$,
COM1: communication to transpose V^{col} into V^{row},
CAL3: $W_2^{\mathrm{row}} := V^{\mathrm{row}} A^T$,
COM2: communication to transpose W_2^{row} into W_2^{col},
CAL4: $W^{\mathrm{col}} := W_1^{\mathrm{col}} + W_2^{\mathrm{col}}$,

where the superscripts 'col' and 'row' denote column-wise and row-wise partitioning, respectively.

The above operational procedure includes the matrix transpose twice which normally requires all-to-all data communication. In the MPI standards, the all-to-all data communication is realized by a collective communication function `MPI_Alltoallv`. However, due to irregular and incontiguous structure of the transferring data, furthermore strong requirement of a non-blocking property (see following subsection), this communication must be composed of a point-to-point or a one-side communication function. Probably it may sound funny that `MPI_Put` is recommended by the developers [12]. However, the one-side communication function `MPI_Put` works more excellently than the point-to-point communication on the ES.

4.3 Communication Overlap

The MPI standard formally guarantees simultaneous execution of computation and communication when it uses the non-blocking point-to-point communications and the one-side communications. This principally enables to hide the communication time behind the computation time, and it is strongly believed that this improves the performance. However, the overlap between communication and computation practically depends on an implementation of the MPI library. In fact, the MPI library installed on the ES had not provided any functions of the overlap until the end of March 2005, and the non-blocking `MPI_Put` had worked as a blocking communication like `MPI_Send`. In the procedure of the matrix-vector multiplication in Sect. 4.2, the calculations CAL1 and CAL2 and the communication COM1 are clearly found to be independently executed. Moreover, although the relation between CAL3 and COM2 is not so simple, the concurrent work can be realized in a pipelining fashion as shown in Fig. 2. Thus, the two communication processes can be potentially hidden behind the calculations.

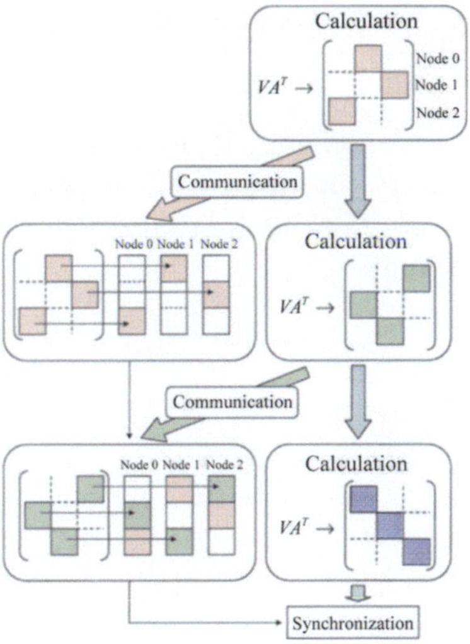

Fig. 2. A data-transfer diagram to overlap VA^T (CAL3) with communication (COM2) in a case using three nodes

As mentioned in previous paragraph, `MPI_Put` installed on the ES prior to the version March 2005 does not work as the non-blocking function[4]. In implementation of our matrix-vector multiplication using the non-blocking `MPI_Put` function, call of `MPI_Win_Fence` to synchronize all processes is required in each pipeline stage. Otherwise, two N-word communication buffers (for send and receive) should be retained until the completion of all the stages. On the other hand, the completion of each stage is assured by return of the `MPI_Put` in the blocking mode, and send-buffer can be repeatedly used. Consequently, one N-word communication buffer becomes free. Thus, we can adopt the blocking `MPI_Put` to extend the maximum limit of the matrix size.

At a glance, this choice seems to sacrifice the overlap functionality of the MPI library. However, one can manage to overlap computation with communication even in the use of the blocking `MPI_Put` on the ES. The way is as follows: The blocking `MPI_Put` can be assigned to a single PE per node by the intra-node parallelization technique. Then, the assigned processor dedicates only the communication task. Consequently, the calculation load is divided into seven PE's. This parallelization strategy, which we call task assignment (TA) method, imitates a non-blocking communication operation, and enables us to overlap the blocking communication with calculation on the ES.

4.4 Effective Usage of Vector Pipelines, and Thread Parallelism

The theoretical FLOPS rate, F, in a single processor of the ES is calculated by

$$F = \frac{4(\#\mathrm{ADD} + \#\mathrm{MUL})}{\max\{\#\mathrm{ADD}, \#\mathrm{MUL}, \#\mathrm{VLD} + \#\mathrm{VST}\}} \text{ GFLOPS}, \tag{3}$$

[4] The latest version supports both non-blocking and blocking modes.

where #ADD, #MUL, #VLD, #VST denote the number of additions, multiplications, vector load, and store operations, respectively. According to the formula (3), the performance of the matrix multiplications AV and VA^T, described in the previous section is normally 2.67 GFLOPS. However, higher order loop unrolling decreases the number of VLD and VST instructions, and improves the performance. In fact, when the degree of loop unrolling is 12 in the multiplication, the performance is estimated to be 6.86 GFLOPS. Moreover,

- the loop fusion,
- the loop reconstruction,
- the efficient and novel vectorizing algorithms [13, 14],
- introduction of explicitly privatized variables (Fig. 3), and so on

improve the single node performance further.

4.5 Performance Estimation

In this section, we estimate the communication overhead and overall performance of our eigenvalue solver.

First, let us summarize the notation of some variables. N basically means the dimension of the system, however, in the matrix-representation the dimension of matrix V becomes $\sqrt{N}$. P is the number of nodes, and in case of the ES each node has 8 PE's. In addition, data type is double precision floating point number, and data size of a single word is 8 Byte.

As presented in previous sections, the core part of our code is the matrix-vector multiplication in both the Lanczos and the PCG methods. We estimate the message size issued on each node in the matrix-vector multiplication as $8N/P^2$ [Byte]. From other work [12] which reports the network performance of the ES, sustained throughput should be assumed 10[GB/s]. Since data communication is carried $2P$ times, therefore, the estimated communication overhead can be calculated $2P \times (8N/P^2[\text{Byte}])/(10[\text{GB/s}]) = 1.6N/P[\text{nsec}]$. Next, we estimate the computational cost. In the matrix-vector multiplication, about $40N/P$ flops are required on each node, and if sustained computational power attains 8×6.8 [GFLOPS] (85% of the peak), the computational cost is estimated

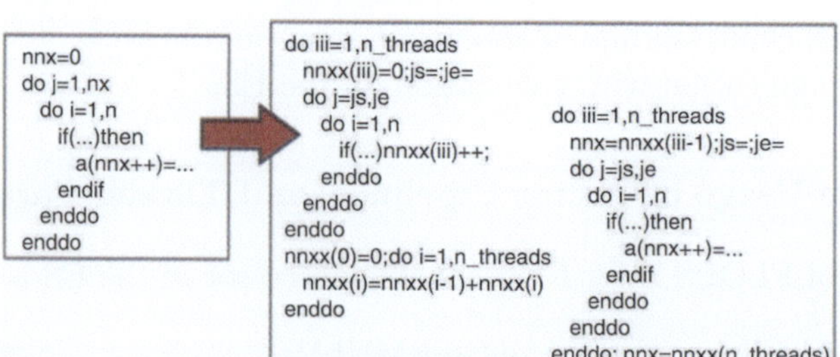

Fig. 3. An example code of loop reconstruction by introducing an explicitly privatized variable. The modified code removes the loop-carried dependency of the variable `nnx`

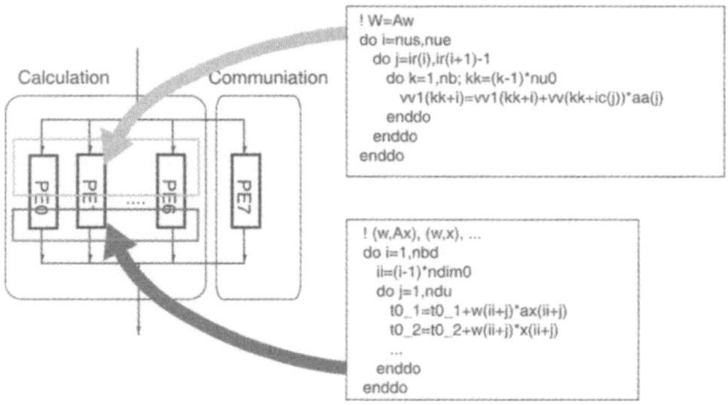

Fig. 4. More effective communication hiding technique, overlapping much more vector operations with communication on our TA method

$(40N/P[\text{flops}])/(8 \times 6.8[\text{GFLOPS}]) = 0.73N/P$ [nsec]. The estimated computational time is equivalent to almost half of the communication overhead, and it suggests the peak performance of the Lanczos method, which considers no effect from other linear algebra parts, is only less than 40% of the peak performance of the ES (at the most 13.10TFLOPS on 512 nodes).

In order to reduce much more communication overhead, we concentrate on concealing communication behind the large amounts of calculations by reordering the vector- and matrix-operations. As shown in Fig. 1(a), the Lanczos method has strong dependency among vector- and matrix-operations, thus, we can not find independent operations further. On the other hand, the PCG method consists of a lot of vector operations, and some of them can work independently, for example, inner-product (not including the term of W_i) can perform with the matrix-vector multiplications in parallel (see Fig. 4). In a rough estimation, $21N/P$ [flops] can be overlapped on each computational node, and half of the idling time is removed from our code.

In deed, some results presented in previous sections apply the communication hiding techniques shown here. One can easily understand that the performance results of the PCG demonstrate the effect of reducing the communication overhead. In Sect. 5, we examine our eigensolver on a lager partition on the ES, 512 nodes, which is the largest partition opened for non-administrative users.

5 Performance on the Earth Simulator

The performance of the Lanczos method and the PCG method with the TA method for huge Hamiltonian matrices is presented in Table 3 and 4. Table 3 shows the system configurations, specifically, the numbers of sites and fermions and the matrix dimension. Table 4 shows the performance of these methods on 512 nodes of the ES.

The total elapsed time and FLOPS rates are measured by using the built-in performance analysis routine [15] installed on the ES. On the other hand, the FLOPS rates of the solvers are evaluated by the elapsed time and the flops count summed up by hand (the ratio of the computational cost per iteration

Table 3. The dimension of Hamiltonian matrix H, the number of nodes, and memory requirements. In case of the model 1 on the PCG method, memory requirement is beyond 10TB

Model	No. of Sites	No. of Fermions (↑ / ↓ spin)	Dimension of H	No. of Nodes	Memory [TB] Lanczos	Memory [TB] PCG
1	24	7/7	119,787,978,816	512	7.0	na
2	22	8/8	102,252,852,900	512	4.6	6.9

Table 4. Performances of the Lanczos method and the PCG method on the ES (March 2005)

Model	Lanczos method				PCG method			
	Itr.	Residual Error	Elapsed time [sec] Total	Elapsed time [sec] Solver	Itr.	Residual Error	Elapsed time [sec] Total	Elapsed time [sec] Solver
1 (TFLOPS)	200	5.4E-8	233.849 (10.215)	173.355 (11.170)	–	–	– –	– –
2 (TFLOPS)	300	3.6E-11	288.270 (10.613)	279.775 (10.906)	109	2.4E-9	68.079 (14.500)	60.640 (16.140)

between the Lanczos and the PCG is roughly 2:3). As shown in Table 4, the PCG method shows better convergence property, and it solves the eigenvalue problems less than one third iteration of the Lanczos method. Moreover, concerning the ratio between the elapsed time and flops count of both methods, the PCG method performs excellently. It can be interpreted that the PCG method overlaps communication with calculations much more effectively.

The best performance of the PCG method is 16.14TFLOPS on 512 nodes which is 49.3% of the theoretical peak. On the other hand, Table 3 and 4 show that the Lanczos method can solve up to the 120-billion-dimensional Hamiltonian matrix on 512 nodes. To our knowledge, this size is the largest in the history of the exact diagonalization method of Hamiltonian matrices.

6 Conclusions

The best performance, 16.14TFLOPS, of our high performance eigensolver is comparable to those of other applications on the Earth Simulator as reported in the Supercomputing conferences. However, we would like to point out that our application requires massive communications in contrast to the previous ones. We made many efforts to reduce the communication overhead by paying an attention to the architecture of the Earth Simulator. As a result, we confirmed that the PCG method shows the best performance, and drastically shorten the total elapsed time. This is quite useful for systematic calculations like the present simulation code. The best performance by the PCG method and the world record of the large matrix operation are achieved. We believe that these results contribute to not only Tera-FLOPS computing but also the next step of HPC, Peta-FLOPS computing.

Acknowledgements

The authors would like to thank G. Yagawa, T. Hirayama, C. Arakawa, N. Inoue and T. Kano for their supports, and acknowledge K. Itakura and staff members in the Earth Simulator Center of JAMSTEC for their supports in the present calculations. One of the authors, M.M., acknowledges T. Egami and P. Piekarz for illuminating discussion about diagonalization for d-p model and H. Matsumoto and Y. Ohashi for their collaboration on the optical-lattice fermion systems.

References

1. Machida M., Yamada S., Ohashi Y., Matsumoto H.: Novel Superfluidity in a Trapped Gas of Fermi Atoms with Repulsive Interaction Loaded on an Optical Lattice. Phys. Rev. Lett., **93** (2004) 200402
2. Rasetti M. (ed.): The Hubbard Model: Recent Results. Series on Advances in Statistical Mechanics, Vol. 7., World Scientific, Singapore (1991)
3. Montorsi A. (ed.): The Hubbard Model: A Collection of Reprints. World Scientific, Singapore (1992)
4. Rigol M., Muramatsu A., Batrouni G.G., Scalettar R.T.: Local Quantum Criticality in Confined Fermions on Optical Lattices. Phys. Rev. Lett., **91** (2003) 130403
5. Dagotto E.: Correlated Electrons in High-temperature Superconductors. Rev. Mod. Phys., **66** (1994) 763
6. The Earth Simulator Center. `http://www.es.jamstec.go.jp/esc/eng/`
7. TOP500 Supercomputer Sites. `http://www.top500.org/`
8. Shingu S. et al.: A 26.58 Tflops Global Atmospheric Simulation with the Spectral Transform Method on the Earth Simulator. Proc. of SC2002, IEEE/ACM (2002)
9. Yamada S., Imamura T., Machida M.: 10TFLOPS Eigenvalue Solver for Strongly-Correlated Fermions on the Earth Simulator. Proc. of PDCN2005, IASTED (2005)
10. Cullum J.K., Willoughby R.A.: Lanczos Algorithms for Large Symmetric Eigenvalue Computations, Vol. 1. SIAM, Philadelphia PA (2002)
11. Knyazev A.V.: Preconditioned Eigensolvers – An Oxymoron? Electr. Trans. on Numer. Anal., Vol. 7 (1998) 104–123
12. Uehara H., Tamura M., Yokokawa M.: MPI Performance Measurement on the Earth Simulator. NEC Research & Development, Vol. 44, No. 1 (2003) 75–79
13. Vorst H.A., Dekker K.: Vectorization of Linear Recurrence Relations. SIAM J. Sci. Stat. Comput., Vol. 10, No. 1 (1989) 27–35
14. Imamura T.: A Group of Retry-type Algorithms on a Vector Computer. IPSJ, Trans., Vol. 46, SIG 7 (2005) 52–62 (*written in Japanese*)
15. NEC Corporation, FORTRAN90/ES Programmerfs Guide, Earth Simulator Userfs Manuals. NEC Corporation (2002)

First-Principles Simulation on Femtosecond Dynamics in Condensed Matters Within TDDFT-MD Approach

Yoshiyuki Miyamoto*

Fundamental and Environmental Research Laboratories, NEC Corp.,
34 Miyukigaoka, Tsukuba, 305-8501, Japan,
y-miyamoto@ce.jp.nec.com

Abstract In this article, we introduce a new approach based on the *time-dependent* density functional theory (TDDFT), where the *real-time propagation* of the Kohn-Sham wave functions of electrons are treated by integrating the time-evolution operator. We have combined this technique with conventional classical molecular dynamics simulation for ions in order to see very fast phenomena in condensed matters like as photo-induced chemical reactions and hot-carrier dynamics. We briefly introduce this technique and demonstrate some examples of ultra-fast phenomena in carbon nanotubes.

1 Introduction

In 1999, Professor Ahmed H. Zewail received the Nobel Prize in Chemistry for his studies on transition states of chemical reaction using the femtosecond spectroscopy. (1 femtosecond (fs) = 10^{-15} seconds.) This technique opened a door to very fast phenomena in the typical time constant of hundreds fs. Meanwhile, theoretical methods so-called as *ab initio* or *first-principles* methods, based on *time-independent* Schrödinger equation, are less powerful to understand phenomena within this time regime. This is because the conventional concept of the thermal equilibrium or Fermi-Golden rule does not work and electron-dynamics must be directly treated.

Density functional theory (DFT) [1] enabled us to treat single-particle representation of electron wave functions in condensed matters even with many-

* The author is indebted to Professor Osamu Sugino for his great contribution in developing the computer code "FPSEID" (éf-psái-dí:), which means First-Principles Simulation tool for Electron Ion Dynamics. The MPI version of the FPSEID has been developed with a help of Mr. Takeshi Kurimoto and CCRL MPI-team at NEC Europe (Bonn). The researches on carbon nanotubes were done in collaboration with Professors Angel Rubio and David Tománek. Most of the calculations were performed by using the Earth Simulator with a help by Noboru Jinbo.

body interactions. This is owing to the theorem of one-to-one relationship between the charge density and the Hartree-exchange-correlation potential of electrons. Thanks to this theorem, variational Euler equation of the total-energy turns out to be Kohn-Sham equation [2], which is a DFT version of the *time-independent* Schrödinger equation. Runge and Gross derived the *time-dependent* Kohn-Sham equation [3] from the Euler equation of the "action" by extending the one-to-one relationship into space and time. The usefulness of the *time-dependent* DFT (TDDFT) [3] was demonstrated by Yabana and Bertsch [4], who succeeded to improve the computed optical spectroscopy of finite systems by Fourier-transforming the time-varying dipole moment initiated by a finite replacement of electron clouds.

In this manuscript, we demonstrate that the use of TDDFT combined with the molecular dynamics (MD) simulation is a powerful tool for approaching the ultra-fast phenomena under electronic excitations [5]. In addition to the 'real-time propagation' of electrons [4], we treat ionic motion within Ehrenfest approximation [6]. Since ion dynamics requires typical simulation time in the order of hundreds fs, we need numerical stability in solving the *time-dependent* Schrödinger equation for such a time constant. We chose the *Suzuki-Trotter* split operator method [7], where an accuracy up to fourth order with respect to the time-step *dt* is guaranteed. We believe that our TDDFT-MD simulations will be verified by the pump-probe technique using the femtosecond laser.

The rest of this manuscript is organized as follows: In Sect. 2, we briefly explain how to perform the MD simulation under electronic excitation. In Sect. 3, we present application of TDDFT-MD simulation for optical excitation and subsequent dynamics in carbon nanotubes. We demonstrate two examples. The first one is spontaneous emission of an oxygen (O) impurity atom from carbon nanotube, and the second one is rapid reduction of the energy gap of hot-electron and hot-hole created in carbon nanotubes by optical excitation. In Sect. 4, we summarize and present future aspects of the TDDFT simulations.

2 Computational Methods

In order to perform MD simulation under electronic excitation, electron dynamics on real-time axis must be treated because of following reasons. The excited state at particular atomic configuration can be mimicked by promoting electronic occupation and solving the *time-independent* Schrödinger equation. However, when atomic positions are allowed to move, level alternation among the states with different occupation numbers often occurs. When the *time-independent* Schrödinger equation is used throughout the MD simulation, the level assignment is very hard and sometimes is made with mistake. On the other hand, time-evolution technique by integrating the *time-dependent* Schrödinger equation enables us to know which state in current time originated from which state in the past, so we can proceed MD simulation under the electronic excitation with a substantial numerical stability.

The *time-dependent* Schrödinger equation has a form like,

$$i\hbar\frac{d\psi_n}{dt} = H\psi_n, \tag{1}$$

where $\hbar$ means the Plank constant divided by 2π. H is the Hamiltonian of the system of the interest and ψ_n represents wave function of electron and subscript n means quantum number. When the Hamiltonian H depends on time, the time-integration of Eq. (1) can be written as,

$$\psi_n(t+dt) = \psi_n(t) + \int_t^{t+dt}\left(\frac{-i}{\hbar}\right)H(t_1)\psi_n(t_1)dt_1. \tag{2}$$

Since Eq. (2) still remain unknown function $\psi_n(t_1)$, we should repeat the integration as follows,

$$\begin{aligned}
\psi_n(t+dt) &= \psi_n(t) + \int_t^{t+dt}\left(\frac{-i}{\hbar}\right)H(t_1)\psi_n(t)dt_1 \\
&+ \int_t^{t+dt}\int_t^{t_1}\left(\frac{-i}{\hbar}\right)^2 H(t_1)H(t_2)\psi_n(t_2)dt_2dt_1 \\
&\quad \cdots\cdots \\
&= \psi_n(t) + \sum_{N=1}^{\infty}\left(\frac{-i}{\hbar}\right)^N \int_t^{t+dt}\int_t^{t_1}\cdots \\
&\times \int_t^{t_{N-1}} H(t_1)H(t_2)\cdots H(t_N)\psi_n(t_N)dt_N\cdots dt_2dt_1 \\
&= \psi_n(t) + T\sum_{N=1}^{\infty}\frac{1}{N!}\left(\frac{-i}{\hbar}\right)^N \int_t^{t+dt}\int_t^{t+dt}\cdots \\
&\times \int_t^{t+dt} H(t_1)H(t_2)\cdots H(t_N)\psi_n(t_N)dt_N\cdots dt_2dt_1 \\
&= T\int_t^{t+dt} e^{\frac{-i}{\hbar}H(t')}\psi_n(t')dt',
\end{aligned} \tag{3}$$

where, T means the time-reordering operator.

In a practical sense, performing multiple integral along with the time-axis is not feasible. We therefore use time-evolution scheme like as,

$$\psi_n(t+dt) = e^{\frac{-i}{\hbar}H}\psi_n(t), \tag{4}$$

making dt so small as to keep the numerical accuracy.

Now, we move on the first-principles calculation based on the DFT with use of the pseudopotentials to express interactions between valence electrons and ions. Generally, the pseudopotentials contain non-local operation and thus the Hamiltonian H can be written as,

$$H = -\frac{\hbar^2}{2m}\nabla^2 + \sum_{\tau,(l,m)} V_{nl}(\tau; l.m) + V_{HXC}(\mathbf{r}, t), \tag{5}$$

where the first term is a kinetic energy operator for electrons and the last term is local Hartree-exchange-correlation potential, which is a functional of the charge density in the DFT. The middle is a summation of the non-local parts of the pseudopotentials with atomic site τ and angular quantum numbers l and m, which includes information of atomic pseudo-wave-functions at site τ. The local-part of the pseudopotentials is not clearly written here, which should be effectively included to local potential $V_{HXC}(\mathbf{r}, t)$, with $\mathbf{r}$ as a coordinate of electron.

Note that all operators included in Eq. (5) do not commute to each other and the number of the operators (non-local terms of the pseudopotentials) depends on how many atoms are considered in the system of interest. It is therefore rather complicated to consider the exponential of Eq. (5) compared to the simple Trotter-scheme where H contains two terms only, *i.e.*, the kinetic energy term and the local potential term. However, Suzuki [7] discovered general rule to express the exponential of $H = A_1 + A_2 + \cdots + A_{q-1} + A_q$ like follows,

$$e^{xH} \sim e^{\frac{x}{2}A_1} e^{\frac{x}{2}A_2} \cdots e^{\frac{x}{2}A_{q-1}} e^{xA_q} e^{\frac{x}{2}A_{q-1}} \cdots e^{\frac{x}{2}A_2} e^{\frac{x}{2}A_1} \equiv S_2(x), \tag{6}$$

where x is $\frac{-i}{\hbar}dt$. Here, of course, operators $A_1, A_2, \cdots A_{q-1}, A_q$ individually corresponds to terms of Eq. (5).

Furthermore, Suzuki [7] found that higher order of accuracy can be achieved by repeatedly operating $S_2(x)$ as,

$$S_4(x) \equiv S_2(P_1 x) S_2(P_2 x) S_2(P_3 x) S_2(P_2 x) S_2(P_1 x), \tag{7}$$

where

$$P_1 = P_2 = \frac{1}{4 - 4^{1/3}},$$
$$P_3 = 1 - 4P_1. \tag{8}$$

We have tested expressions with further higher-orders [7] and found that the fourth order expression (Eq. (7)) is accurate enough for our numerical simulation [5] based on the TDDFT [3].

Since we are now able to split the time-evolution (Eq. (4)) into a series of exponential of individual operators shown in Eq. (5), next thing we should focus on is how to proceed the operation of each exponential to the Kohn-Sham wave functions. Here, we consider plane wave basis set scheme with which the Kohn-Sham wave function $\psi_n(\mathbf{r}, t)$ can be expressed by,

$$\psi_n(\mathbf{r}, t) = \frac{1}{\sqrt{\Omega}} \sum_{\mathbf{G}} C^n_{\mathbf{G},\mathbf{K}}(t) e^{i(\mathbf{G}+\mathbf{k})\cdot\mathbf{r}}. \tag{9}$$

Here Ω is a volume of unit cell used for the band-structure calculation, $\mathbf{G}$ and $\mathbf{k}$ are reciprocal and Bloch vectors.

When we operate exponential of the kinetic energy operator shown in Eq. (5), the operation can be directly done in the reciprocal space using the right hand side of the Eq. (9) as,

$$e^{\left(\frac{-i}{\hbar}dt\right)\frac{-\hbar^2}{2m}\nabla}\psi_n(\mathbf{r},t) = \frac{1}{\sqrt{\Omega}}\sum_{\mathbf{G}} e^{\left(\frac{-i}{\hbar}dt\right)\frac{\hbar^2}{2m}(\mathbf{G}+\mathbf{k})^2} C^n_{\mathbf{G},\mathbf{k}}(t)e^{i(\mathbf{G}+\mathbf{k})\cdot\mathbf{r}}. \tag{10}$$

On the other hand, the exponential of the local potential $V_{HXC}(\mathbf{r},t)$ can directly operate to $\psi_n(\mathbf{r},t)$ in real space as,

$$e^{\frac{-i}{\hbar}V_{HXC}(\mathbf{r},t)}\psi_n(\mathbf{r},t), \tag{11}$$

The exponential of the non-local part seems to be rather complicated. Yet if the non-local term has a separable form [8] like,

$$V_{nl}(\tau;l.m) = \frac{V_{rad}(l,m) \mid \phi^{l.m}_{ps}(\tau)\rangle\langle\phi^{l.m}_{ps}(\tau) \mid V_{rad}(l,m)}{\langle\phi^{l.m}_{ps} \mid V^{l,m}_{rad} \mid \phi^{l.m}_{ps}\rangle}. \tag{12}$$

the treatment becomes straightforward. Here $\phi^{l.m}_{ps}$ means atomic pseudo wave functions with a set of angular quantum numbers l and m, and $V^{l,m}_{rad}$ is spherical potential.

Multiple operations of the operator of Eq. (12) can easily obtained as,

$$V_{nl}(\tau;l,m)^N = \frac{V^{l,m}_{rad} \mid \phi^{l,m}_{ps}(\tau)\rangle\langle\phi^{l,m}_{ps} \mid (V^{l,m}_{rad})^2 \mid \phi^{l,m}_{ps}\rangle^{N-1}\langle\phi^{l,m}_{ps}(\tau) \mid V^{l,m}_{rad}}{\langle\phi^{l.m}_{ps} \mid V^{l,m}_{rad} \mid \phi^{l.m}_{ps}\rangle^N}, \tag{13}$$

with $N > 0$.

The Eq. (13) can simply be used to express infinite Taylor expansion of an exponential of the non-local part as,

$$e^{xV_{nl}(\tau;l,m)} = \sum_{N=0}^{\infty}\frac{1}{N!}V_{nl}(\tau;l,m)^N$$

$$= 1 + V^{l,m}_{rad} \mid \phi^{l,m}_{ps}(\tau)\rangle\left(e^{x\frac{\langle\phi^{l,m}_{ps}|(V^{l,m}_{rad})^2|\phi^{l,m}_{ps}\rangle}{\langle\phi^{l,m}_{ps}|V^{l,m}_{rad}|\phi^{l,m}_{ps}\rangle}} - 1\right)/\langle\phi^{l,m}_{ps} \mid (V^{l,m}_{rad})^2 \mid \phi^{l,m}_{ps}\rangle$$

$$\times\langle\phi^{l,m}_{ps}(\tau) \mid V^{l,m}_{rad}, \tag{14}$$

with $x = \frac{-i}{\hbar}dt$. Equation (14) shows that operation of an exponential of the non-local part of the pseudopotential can be done in the same manner as the operation of the original pseudopotentials.

To proceed integration of the time-evolution operator (Eq. (4)), we repeatedly operate exponentials of each operator included in the Kohn-Sham Hamiltonian (Eq. (5)). Fast Fourier Transformation (FFT) is used to convert wave functions from reciprocal space to real space just before operating the exponential of the local potential (Eq. (11)), then the wave functions are re-converted into reciprocal space to proceed operations of Eq. (10) and Eq. (14). Of course one can do operation of Eq. (14) in real space, too. Unlike to the conventional plane-wave-band-structure calculations, we need to use full-grid for the FFT in the reciprocal space in order to avoid numerical noise throughout the simulation [5]. This fact requires larger core-memory of processor than conventional band-structure calculations.

Usage of the split-operator method automatically keeps the ortho-normal condition of the set of the wave functions. This is a big advantage for the parallel computing, where each processor can share the task for the time-propagation of wave functions without communicating to each other.

In addition to the split-operator methods, we apply further technique to reach the stability, the details of which are described in our former report [5]. We have developed a parallelized code FPSEID (éf-psái-dí:), which means First-Principles Simulation tool for Electron Ion Dynamics. Although the program is well parallelized and suitable for large systems, we cannot parallelize the calculation along with the time-axis due to the 'causality' principle. We therefore still require speed of each processor for simulating the long-time phenomena.

The procedure of our calculation for the excited state MD simulation is as followings: First we perform conventional band-structure and total-energy calculations, and perform the geometry optimization according to the computed forces on ions. Then we artificially promote the occupation numbers of electronic states to mimic the excited states as mentioned in the beginning of this section. We analyze characteristics of each wave function to search possible excitation pair obtained by optical-dipole transition. After reaching the condition of the self-consistent field (SCF) between $V_{HXC}(\mathbf{r},t)$ and the charge density, we start the time-evolution of the wave functions and the MD simulations. Throughout the TDDFT-MD simulation, we keep the self-consistency between the $V_{HXC}(\mathbf{r},t)$ and the time-evolving charge density made by a sum of norm of the time-evolving Kohn-Sham wave functions $|\psi_n(\mathbf{r},t)|^2$. We have experienced stability of the simulation, which can be confirmed by conservation of the total energy, *i.e.*, the potential energy plus kinetic energy of ions. Even when the simulation time is beyond pico-second, we don't see initiation of the instability of the simulation. As far as we know, such stability cannot be seen in other TDDFT-MD simulations with *real-time* propagation.

3 Application of TDDFT-MD Simulation

In this section, we describe application of the TDDFT-MD simulations to excited-state dynamics in carbon nanotubes. Carbon nanotubes have attracted many attentions from both scientific and technological viewpoints because of their variety of chirality [9] and significant toughness despite their small diameters. The application of nanotube in electronic devices yet has a lot of hurdles since the intrinsic impurities and carrier dynamics are not clearly known.

We explore possible structure of the O impurities in nanotube and propose an efficient method to safely remove them without destructing the remaining C-C-bond network. We also investigate mechanisms of the hot-carrier decay in very short time-constant, which can be divided into the two time-domains for electron-electron coupling and electron-phonon coupling.

All calculations reported here were performed by using a functional form for the exchange-correlation potential [10] fitted to the numerical calculation [11]. As for the pseudopotentials, we adopted norm-conserving pseudopotentials [12].

By performing the force calculation, we follow the scheme of the total energy and force calculation in periodic systems [13]. The cutoff energy for the plane wave basis set in expressing the wave functions are 60 Ry and 40 Ry for cases with and without O atoms, respectively.

3.1 Removal of O Impurities from Carbon Nanotubes

The most widely used method for growing the carbon nanotube is the chemical vapor deposition (CVD) technique which can fabricate carbon nanotubes on patterned substrates [14]. The CVD method requires introduction of either alcohol [15] or water [16] in addition to the source gases like as methane (CH_4) or ethylene (C_2H_2). This condition is necessary to remove amorphous carbon, but causes a risk of contamination by O impurities. The presence of O impurities is inferred by near edge X-ray absorption fine structure spectroscopy [17], which suggests formation of chemically strong C-O-C complexes in the C-C honeycomb network of the carbon nanotubes. If this is the case, removal of O impurity by thermal processes inevitably hurts the C-C-bonds which is manifest from emission of CO and CO_2 molecules with increased temperature [18].

The left panel of Fig. 1 shows possible structure of an O impurity atom in carbon nanotube making C-O-C complex. In this geometry all C atoms are three-hold coordinated so there are no dangling bonds. The system is chemically stable and the O atom is hard to be removed even when a radical hydrogen atom attacks the O atom. Despite the chemical stability of the C-O-C complex, this

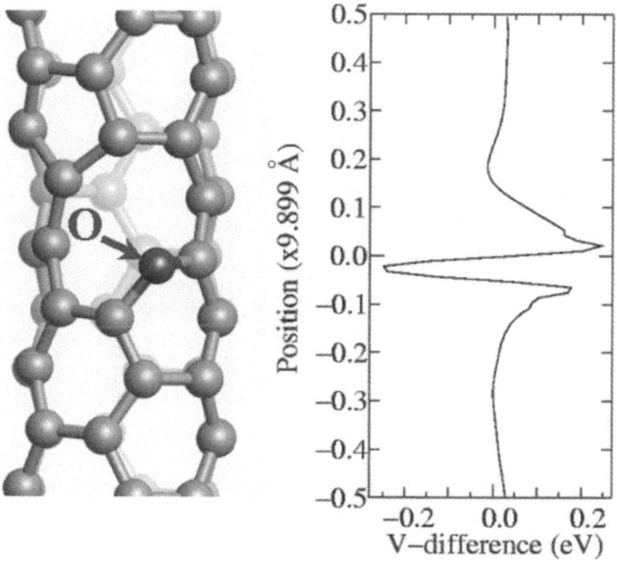

Fig. 1. (left) O impurity atom in a (3,3) nanotube and (right) corresponding SCF potential profile which is obtained by taking difference of the SCF potentials with and without O impurity. The potential is averaged in directions perpendicular to tube axis

complex disturbs conduction of electron through this carbon nanotube. The right panel of Fig. 1 shows modification of the self-consistent potential for electron due to presence of the O impurity. One can note existence of hump and dip of the potential along with the tube axis which causes either scattering or trapping of conducting carriers. Therefore, removal of O impurity will be an important technology of carbon nanotubes based devices.

The electronic structure of C-O-C complex gives us a hint to weaken local C-O-C bond. Below the valence band of the carbon nanotube, there exists highly localized orbital, which is dominated by O $2s$ orbital and be hybridized with $2p$ orbital of neighboring C atom in the bonding phase. Let us call this level as state 'A'. Meanwhile in a resonance of conduction bands of carbon nanotube, another localized state exists. This orbital is dominated by O $2p$ orbital being hybridized with $2p$ orbital of neighboring C atom in the anti-bonding phase. Let us call this empty level as state 'B'. One can therefore expect that photo-excitation from state 'A' to state 'B' can weaken the C-O-C chemical bonds. A schematic diagram for the electronic energy level is shown in Fig. 2 (a).

However, according to our TDDFT-MD simulation, single excitation from state 'A' to state 'B' with corresponding photo-excitation energy of 33 eV does not complete O-emission from carbon nanotube. The O atom shows an oscillation but this motion dissipated into lattice vibration of entire system. (The corresponding snap shots are not shown here.) We therefore change our idea to

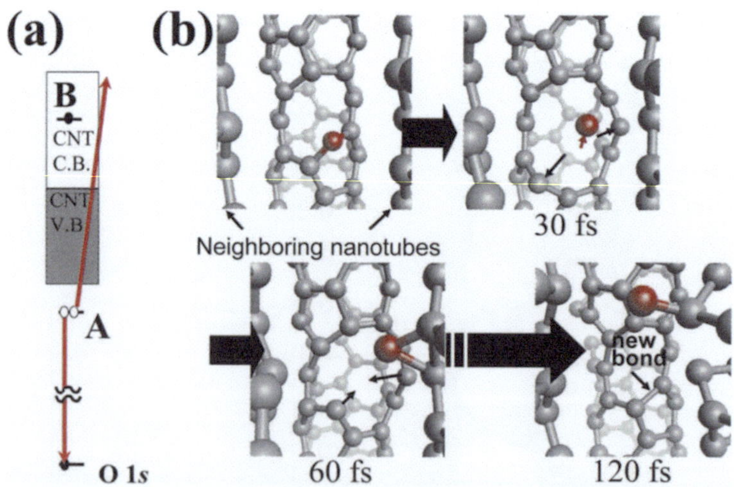

Fig. 2. (**a**) Schematic diagram of electronic structure of C-O-C complex in a carbon nanotubes. Arrows indicate Auger process upon O $1s$ core-excitation into the state 'B'. Two holes in the state 'A' are shown as two open circles. Arrows denote relaxation of one electron of state 'A' into O $1s$ core state, and emission of the other electron of state 'A' into vacuum. (**b**) Snap shots of spontaneous O-emission from carbon nanotubes. Directions of atomic motion of an O atom and its neighbors (C atoms) are also denoted by arrows

excite O 1s core-electron to state 'B' with corresponding excitation energy of 520 eV. This excitation can cause an Auger process remaining two holes in state 'A', as shown in a schematic picture in Fig. 2 (a). We set this Auger final state as the initial condition of our MD simulation and start the TDDFT-MD simulation. The snap shots of the simulation are shown in Fig. 2 (b), which show spontaneous O emission from carbon nanotube. Just after the emission, the neighboring C atoms are kicked out to enlarge the size of remaining vacancy. But as shown in the following snap shots, the carbon nanotube recover its cylindrical shape by forming a new C-C bond like as the final snap shot of Fig. 2 (b).

On the other hand, one can note that the emitted O atom behaves as O-radical, which indeed attacks other side of the carbon nanotube as has been displayed in Fig. 2 (b). To avoid such re-oxidation, we found that introduction of H_2 molecule is effective. H_2 molecule reacts weakly with carbon nanotubes (physisorption), while it reacts strongly with emitted O atom and forms H-O chemical bond before the emitted O atom attacks other side of the carbon nanotube. We therefore conclude that a combination of O 1s core excitation and introduction of H_2 molecule is an efficient method to remove O impurity from carbon nanotubes and would be useful to refine quality of carbon nanotube even after the fabrication of the nanotube-devices. More detailed conditions of the present calculations and results are shown in our former report [19].

3.2 Ultra-Fast Decay of Hot-Carrier in Carbon Nanotubes

Application of carbon nanotubes for high-frequency devices such as transistor [20, 21] and optical limiting switch [22] is a current hot topic. This type of application needs basic understanding of decay dynamics of excited carrier. If the carrier lifetime is too short, we have low quantum efficiency. Meanwhile if the lifetime is too long, the frequency of the device operation should be low. Recently, measurement of carrier dynamics has been made by use of the femtosecond laser [23, 24]. These experiments suggested ultra-fast decay of hot-carriers which can be divided into two-time domains: rapid decay within 200 fs and slower decay ranging over picoseconds. The earlier decay is interpreted as electron-electron coupling while the slower one is interpreted as electron-phonon coupling. These measurements were, however, done with samples containing carbon nanotubes with variety of chiralities. Therefore, the experimental data must be a superposition of decay dynamics with different time-constants. Therefore the intrinsic property of the carbon nanotube is not well understood.

We here perform simulation for dynamics of hot-carriers in an isolated carbon nanotube with a particular chirality. We here assume arm-chair type nanotube with very small diameter of 4 Å, *i.e.*, the (3,3) nanotube. Optical absorption of such thin nanotubes was reported before [25] in energy region from 1 eV to 3 eV. Meanwhile, we found that optical transition in higher energy region is also available in such nanotubes according to our first-principles calculation of dipole-matrix elements. We suspected that such high-energy excitation bores states with very short lifetime and cannot be observed as a recognizable peak in the absorption spectrum.

We promote electronic occupation to make hot hole and hot electron in the (3,3) nanotube with the corresponding excitation energy of 6.8 eV within the local density approximation. We found that an electron-hole pair made by this excitation has non-zero optical matrix elements with dipole-vector parallel to the tube axis. Then we prepare initial lattice velocities with a set of randomized numbers which follows Maxwell-Boltzmann distribution function under the room temperature. With these initial conditions, we started the TDDFT-MD simulation. Throughout the simulation, we do not use the thermostat [26, 27] to allow lattice to be heated up by excited electrons.

Figure 3 shows time-evolution of the single-electron's expectation value,

$$\langle \psi_n(\mathbf{r},t) \mid H(\mathbf{r},t) \mid \psi_n(\mathbf{r},t)\rangle. \tag{15}$$

One can note that the energy gap of hot-electron and hot-hole rapidly reduces less than quarter of the original value within 600 fs. Another significant feature is many events of the level alternation which replace hot-hole and hot-electron in highest occupied and lowest unoccupied level, respectively. Such a massive number of level alternations cannot be dealt with conventional technique solv-

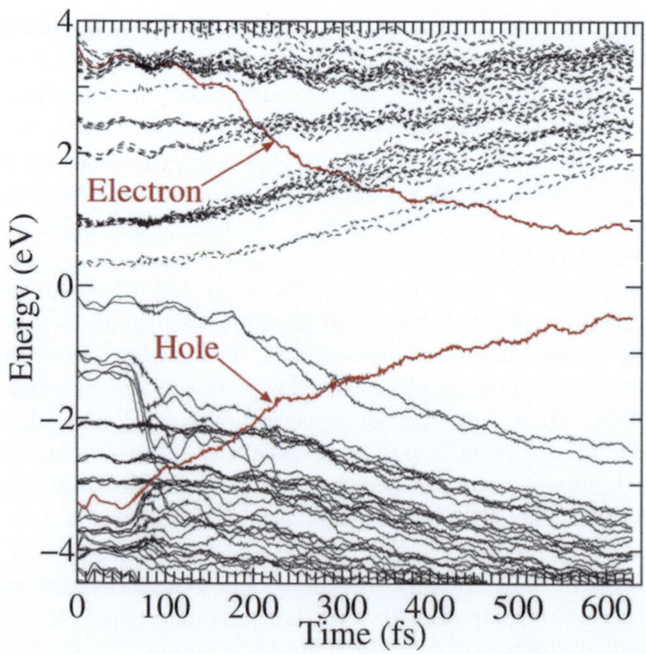

Fig. 3. Time evolution of single-level (expectation values) in a carbon nanotube initiated by photo-excitation by ultraviolet (6.8 eV). The hot-electron and hot-hole are denoted by arrows. Dotted and solid lines are the state in conduction and valence bands, respectively

ing the *time-independent* Schrödinger equation, as mentioned in Sect. 2. The similar level alternations are also seen in the O-emission case in Sect. 3.1. (The corresponding figure is not shown here, but in our former work [19].)

However, only from the data of Fig. 3, it is hard to analyze origin of the decay process mentioned in the introduction of this subsection, *i.e.* electron-electron and electron-phonon couplings. The TDDFT-MD simulation treats both couplings simultaneously, yet we can extract each of them when we see time-evolution of the potential energy for ions. Figure 4 shows the time-evolution of the potential energy throughout the simulation shown in Fig. 3. In the beginning of the simulation there is a large fluctuation of the potential energy, but later the fluctuation becomes less significant. The trend of the time-evolution of the potential can be highlighted when the time-average of the potential is taken according to the following equation,

$$\overline{E_{potential}(t)} = \frac{1}{T} \int_{t-\frac{T}{2}}^{t+\frac{T}{2}} E_{potential}(t')dt', \tag{16}$$

where $E_{potential}(t)$ means potential energy and $\overline{E_{potential}(t)}$ is the one with averaged time T. In Fig. 4, T is set 50 fs. The behavior of $\overline{E_{potential}(t)}$ is rather gentle in the beginning of the simulation while becomes steeper later than 200 fs. The lower drift of the potential means energy transfer from electrons to ions. The steeper slope later than 200 fs means that electron-phonon coupling becomes dominant in that time-regime while electron-electron is rather dominant in earlier time. This is consistent with experimental interpretation [23, 24].

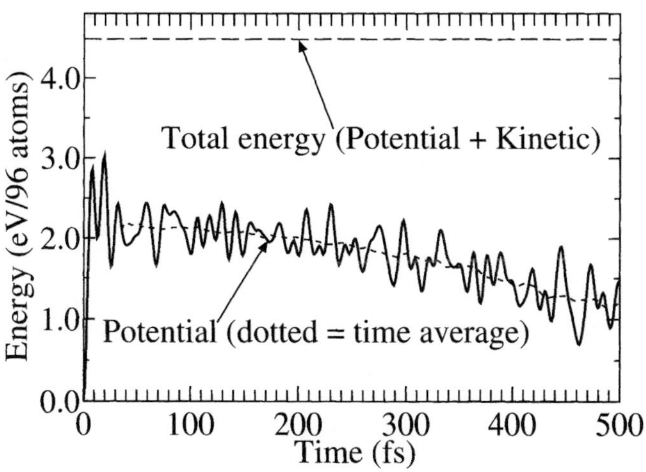

Fig. 4. Time evolution of the total-energy (potential plus kinetic energies shown as a broken line) and potential energy of ions throughout the simulation of Fig. 3. A dotted line is a time-average of the potential energy with the average width of 50 fs

We confirmed our interpretation by performing the similar simulation with deferent initial velocities of ions, which are represented the same set of the randomized numbers with different scales. We found that the turning point on the time-axis from electron-electron coupling to electron-phonon coupling shifts later in slower lattice velocities. We will show the detailed results depending on initial lattice velocities elsewhere.

Since the present decay process is so fast and the system has not reached the thermal equilibrium condition, the conventional Fermi-Golden rule dealing the electron-phonon coupling thus does not work. Under such a non-equilibrium condition, the real-time propagation must be treated like the present TDDFT-MD simulation.

4 Concluding Remarks

We show here feasible first-principles approach on ultra-fast phenomena in condensed matters. By solving the *time-dependent* Schrödinger equation, real-time dynamics of electron can be treated without adjustable parameters. Nevertheless this approach was not performed for real-material simulation because of difficulty in solving the *time-dependent* Schrödinger equation numerically. The split-operator method [7] to the *time-dependent* Schrödinger equation made it able to treat many time-evolving wave functions with an efficient parallel computation and numerical stability. The TDDFT theory has thus become applicable to ultra-fast dynamics of condensed matters under electronic excitation.

The TDDFT-MD simulation will also be applied in the area of bio-materials. For example a time-constant of photoisomerization of retinal is measured as few hundreds fs [28], which would be explained by TDDFT-MD simulation. We expect that mechanism of photo-synthesis, in which transport of excited carrier might be a key factor, will also be solved with an aid of TDDFT-MD simulation.

However, we must note here that the TDDFT-MD method has one difficult problem, when the simulation reached to a point at which different adiabatic potential energy surfaces (PESs) crosses and a probability of non-adiabatic transition grows considerably. There is a traditional quantum chemistry method to practically attack this situation so-called-as 'surface hopping' [29]. But this method is not feasible for extended systems having many PESs. On the other hand, DFT is suitable for extended systems but has an ambiguity at the moment of the non-adiabatic transition. The DFT has Hamiltonian dependent on the charge density, so we must change the Hamiltonian of DFT when the simulation moves from one PES to another one. Change of Hamiltonian means that the wave functions of DFT on different PESs do not belong to the common Hilbert space. This fact makes application of 'surface hopping' basically not appropriate for DFT. Attacking to non-adiabatic transition by TDDFT method is still a challenging problem.

References

1. P. Hohenberg and W. Kohn, Phys. Rev. **136**, B864 (1964).
2. W. Kohn and L. Sham, Phys. Rev. **140**, A1133 (1965).
3. E. Runge and E. K. U Gross, Phys. Rev. Lett. **52**, 997 (1984).
4. K. Yabana and G. F. Bertsch, Phys. Rev. B**54**, 4484 (1996).
5. S. Sugino and Y. Miyamoto, Phys. Rev. B**59**, 2579 (1999); *ibid*, Phys. Rev. B**66** 89901(E) (2002).
6. P. Ehrenfest, Z. Phys. **45**, 455 (1927)
7. M. Suzuki, J. Phys. Soc. Jpn. **61**, L3015 (1992); M. Suzuki and T. Yamauchi, J. Math. Phys. **34**, 4892 (1993).
8. L. Kleinmann and D. M. Bylander, Phys. Rev. Lett. **48**, 1425 (1982).
9. S. Iijima, T. Ichihashi, Nature (London) **363**, 603 (1993).
10. J. P. Perdew, A. Zunger, Phys. Rev. B**23**, 5048 (1981).
11. D. M. Ceperley, B. J. Alder, Phys. Rev. Lett. **45**, 566 (1980).
12. N. Troullier, J. L. Martins, Phys. Rev. B**43**, 1993 (1991).
13. J. Ihm, A. Zunger, and M. L. Cohen, J. Phys. C **12**, 4409 (1979).
14. See, for example, M. Ishida, H. Hongo, F. Nihey, and Y. Ochiai, Jpn. J. Appl. Phys. **43**, L1356 (2004).
15. S. Maruyama, R. Kojima, Y. Miyauchi, S. Chiashi, M. Kohno, Chem. Phys. Lett. **360**, 229 (2002).
16. K. Hata, D. N. Futaba, K. Mizuno, T. Namai M. Yumura, S. Iijima, Science **306**, 1362 (2004).
17. A. Kuznetsova *et al.*, J. Am. Chem. Soc. **123**, 10699 (2001).
18. E. Bekyarova *et al.*, Chem. Phys. Lett. **366**, 463 (200).
19. Y. Miyamoto, N. Jinbo, H. Nakamura, A. Rubio, and D. Tománek, Phys. Rev. B**70**, 233408 (2004).
20. S. Heinze, J. Tersoff, R. Martel, V. Derycke, J. Appenzeller, and Ph. Avouris, Phys. Rev. Lett., **89**, 106801 (2002).
21. F. Nihey, H. Hongo, M. Yudasaka, and S. Iijima, Jpn. J. Appl. Phys. **41**, L1049 (2002).
22. "Mode-locked fiber lasers based on a saturable absorber incorporating carbon nanotubes", by S. Y. Set, H. Yaguchi, M. Jablonski, Y. Tanaka, Y. Sakakibara, A. Rozhin, M. Tokumoto, H. Kataura, Y. Achiba, K. Kikuchi, Proc. of Optical Fiber Communication Conference 2003, March 23–28 (2003).
23. T. Hertel and G. Moos, Phys. Rev. Lett. **84**, 5002 (2000).
24. M. Ichida, Y. Hamanaka, H. Kataura, Y. Achiba, and A. Nakamura, Physica B**323**, 237 (2002).
25. Z. M. Li, *et al.*, Phys. Rev. Lett. **87**, 127401 (2001).
26. S. Nosé, J. Chem. Phys. **81**, 511 (1984).
27. W. G. Hoover, Phys. Rev. A**31**, 1695 (1985).
28. F. Gai, *et. al.*, Science **79**, 1866 (1998).
29. J. C. Tully and R. K. Preston, J. Chem. Phys. **55**, 562 (1971).

Numerical Simulation of Transition and Turbulence in Wall-Bounded Shear Flow

Philipp Schlatter, Steffen Stolz, and Leonhard Kleiser

Institute of Fluid Dynamics, ETH Zurich, 8092 Zurich, Switzerland,
`schlatter@ifd.mavt.ethz.ch`,
WWW home page: `http://www.ifd.mavt.ethz.ch`

Abstract Laminar-turbulent transition encompasses the evolution of a flow from an initially ordered laminar motion into the chaotic turbulent state. Transition is important in a variety of technical applications, however its accurate prediction and the involved physical mechanisms are still a matter of active research.

In the present contribution, an overview is given on recent advances with the simulation of transitional and turbulent incompressible wall-bounded shear flows. The focus is on large-eddy simulations (LES). In LES, only the large-scale, energy-carrying vortices of the flow are accurately resolved on the numerical grid, whereas the small-scale fluctuations, assumed to be more homogeneous, are treated by a subgrid-scale (SGS) model. The application of LES to flows of technical interest is promising as LES provides reasonable accuracy at significantly reduced computational cost compared to fully-resolved direct numerical simulations (DNS). Nevertheless, LES of practical flows still require massive computational resources and the use of supercomputer facilities.

1 Laminar-Turbulent Transition

The behaviour and properties of fluid flows are important in many different technical applications of today's industrial world. One of the most relevant characteristics of a flow is the state in which it is moving: laminar, turbulent, or in the transitional state in between. Laminar flow is well predictable, structured and often stationary, and usually exercises significantly less frictional resistance to solid bodies and much lower mixing rates than the chaotic, swirling and fluctuating state of fluid in turbulent motion. Understanding and predicting both turbulent and transitional flow is crucial in a variety of technical applications, e.g. flows in boundary layers on aircraft wings or around cars, intermittent flows around turbine blades, and flows in chemical reactors or combustion engines. The evolution of an initially laminar flow into a fully developed turbulent flow is referred to as laminar-turbulent transition. This process and specifically the triggering mechanisms of transition are not fully understood even today, after more than a century of research. A summary of developments in transition research is given in the review article by Kachanov (1994) on boundary layer flow and in the recent monograph by Schmid and Henningson (2001).

An overview of laminar-turbulent transition is sketched in Fig. 1 for the canonical case of the flow over a flat plate (boundary-layer transition). The corresponding vortical structures observed during transition in plane channel flow are shown in Fig. 2 (taken from the simulations presented in Schlatter (2005); Schlatter et al. (2006)). The fluid flows along the plate (position ①) until at a certain downstream position, indicated by the Reynolds number Re_{crit}, the laminar flow becomes unstable. Further downstream, two-dimensional wave disturbances grow within the boundary layer (pos. ②) and rapidly evolve into three-dimensional perturbations of triangular shape (Λ-vortices, pos. ③). These vortical structures in turn tend to break down into local turbulent spots through the formation of pronounced hairpin vortices (pos. ④), which grow and merge together to form a fully turbulent boundary layer (pos. ⑤–⑥).

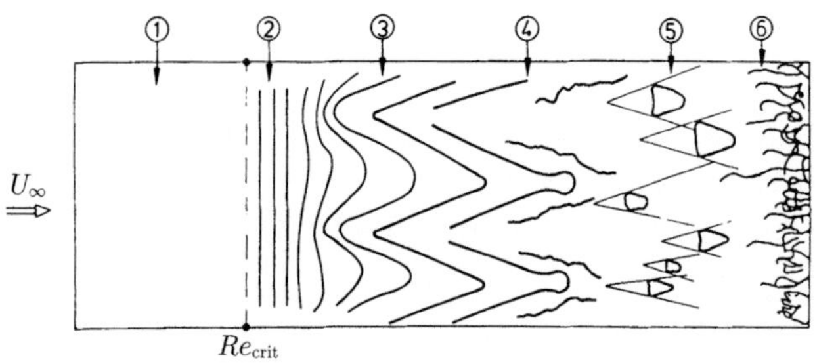

Fig. 1. Schematic view of laminar-turbulent transition in a flat-plate boundary layer (see text for description)

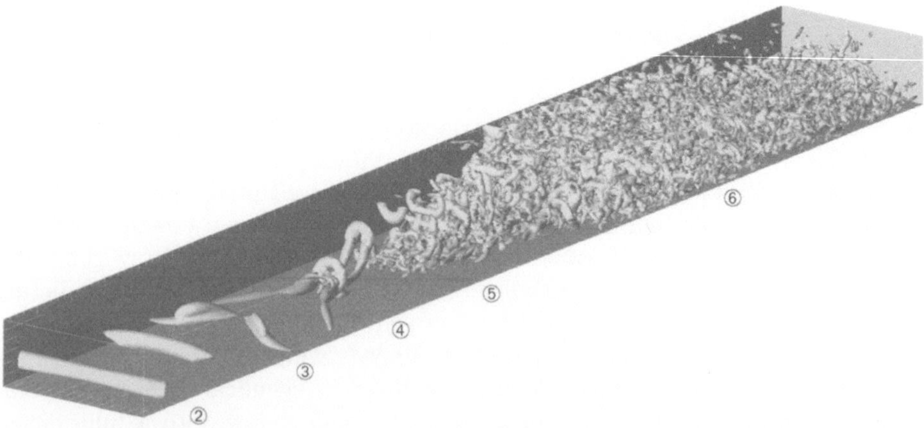

Fig. 2. Visualisation of spatial K-type transition in plane channel flow obtained from a large-eddy simulation (only one channel-half is shown, from Schlatter (2005); Schlatter et al. (2006)). The vortical structures are visualised by the λ_2 criterion (Jeong and Hussain, 1995)

2 Numerical Simulation: DNS and LES

The fully resolved numerical solution of the governing Navier-Stokes equations is referred to as direct numerical simulation (DNS, see e.g. the review by Moin and Mahesh (1998)). In general, it is extremely expensive even for moderate Reynolds numbers Re since the required CPU time roughly scales as Re^3. Practical high Reynolds-number calculations thus need to be performed using simplified turbulence models. The most commonly used possibility is to solve the Reynolds-averaged Navier-Stokes equations (RANS) in which only the mean flow is computed and the effect of the turbulent fluctuations is accounted for by a statistical turbulence model. Although this technique may require a number of empirical ad-hoc adjustments of the turbulence model to a particular flow situation, quite satisfactory results can often be obtained for practical applications.

2.1 Large-Eddy Simulation

A technique with a level of generality in between DNS and RANS is the large-eddy simulation (LES). In an LES, the eddies (turbulent vortices) above a certain size are completely resolved in space and time on the numerical grid, whereas the effect of the smaller scales needs to be modelled. The idea behind this scale separation is that the smaller eddies are more homogeneous and isotropic than the large ones and depend less on the specific flow situation, whereas the energy-carrying large-scale vortices are strongly affected by the particular flow conditions (geometry, inflow, etc.). Since in an LES not all scales have to be resolved on the computational grid, only a fraction of the computational cost compared to fully resolved DNS (typically of order 0.1–1%) is required.

It is expected that LES will play a major role in the future for prediction and analysis of certain complex turbulent flows in which a representation of unsteady turbulent fluctuations is important, such as laminar-turbulent transition, large-scale flow separation in aerodynamics, coupled fluid-structure interaction, turbulent flow control, aeroacoustics and turbulent combustion. However, LES applied to complete configurations (e.g. airplanes) at high Reynolds numbers is still out of reach due to the immense computational effort required by the fine resolution necessary to resolve the turbulent boundary layers.

The success of an LES is essentially dependent on the quality of the underlying subgrid scale (SGS) model and the applied numerical solution scheme. The most prominent SGS model is the Smagorinsky model, which is based on the eddy-viscosity concept and was introduced by Smagorinsky (1963). Substantial research efforts during the past 30 years have led to more universal SGS models. A major generalisation of SGS modelling was achieved by Germano et al. (1991) who proposed an algorithm which allows for dynamically adjusting the model coefficient, a constant to be chosen in the standard Smagorinsky model, to the local flow conditions. In this way the necessary reduction of the model contribution e.g. in the vicinity of walls or in laminar or transitional flow regions is achieved by the model directly (rather than being imposed artificially on an empirical basis).

A different class of SGS models has been introduced by Bardina et al. (1980) (see the review by Meneveau and Katz (2000)) based on the scale-similarity assumption. As the eddy-viscosity closure assumes a one-to-one correlation between the SGS stresses and the large-scale strain rate, the scale-similarity model (SSM) is based upon the idea that the important interactions between the resolved and subgrid scales involve the smallest resolved eddies and the largest SGS eddies.

Considerable research effort has recently been devoted to the development of SGS models of velocity estimation or deconvolution type, see e.g. the review by Domaradzki and Adams (2002). These models can be considered as a generalisation of the scale-similarity approach. An example of such models is the approximate deconvolution model (ADM) developed by Stolz and Adams (1999). ADM has been applied successfully to a number of compressible and incompressible cases (early results see e.g. Stolz et al. (2001a,b)). With the deconvolution-type models, it is tried to extract information about the SGS stresses from the resolved field, thus providing a better approximation of the unknown model terms. Reviews of different strategies for LES and SGS modelling are given in Lesieur and Métais (1996); Domaradzki and Adams (2002); Meneveau and Katz (2000); Piomelli (2001) and in the recent text books by Sagaut (2005) and Geurts (2004).

2.2 Simulation of Transitional Flows

Transitional flows have been the subject of intense experimental and numerical research for many decades. Since the beginning of the 1980s, with the increasing power of computers and reliability and efficiency of numerical algorithms, several researchers have considered the simulation of the breakdown to turbulence in simple incompressible shear flows. One of the first well-resolved simulations to actually compute three-dimensional transition and the following fully developed turbulence was presented by Gilbert and Kleiser (1990), who simulated fundamental K-type transition in plane Poiseuille flow. Comprehensive review articles on the numerical simulation of transition can be found in Kleiser and Zang (1991) and Rempfer (2003).

In transitional flows one is typically dealing with stability problems where small initial disturbances with energies many orders of magnitude smaller than the energy of the steady base flow are amplified and may finally evolve into turbulent fluctuations. After disturbance growth and breakdown the resulting energy of the turbulent fluctuations may be nearly of the same order as that of the base flow. Moreover, the spatial and temporal evolution of various wave disturbances and their nonlinear interaction needs to be computed accurately over many disturbance cycles. These specific challenges have to be addressed if one attempts to accurately simulate laminar-turbulent transition and make this task one of the most demanding ones of computational fluid dynamics.

An SGS model suitable to simulate transition should be able to deal equally well with laminar, various stages of transitional, and turbulent flow states. The model should leave the laminar base flow unaffected and only be effective, in an appropriate way, when and where interactions between the resolved modes and the non-resolved scales become important. The initial slow growth of instability waves is usually sufficiently well resolved even on a coarse LES grid.

3 LES of Transitional Flows

While a number of different LES subgrid-scale models with applications to turbulent flows have been reported in the literature (see the reviews mentioned above), the application of SGS models to transitional flows has become an active field of research only recently. Nevertheless, a number of successful applications of LES to transitional flows are available, most of them based on an eddy-viscosity assumption using a variant of the Smagorinsky model.

3.1 Previous Work

It is well known that the Smagorinsky model in its original formulation is too dissipative and usually, supplementary to distorting laminar flows, relaminarises transitional flows. Consequently, Piomelli et al. (1990) introduced, in addition to the van Driest wall-damping function (van Driest, 1956), an intermittency correction in the eddy-viscosity to decrease the dissipation in (nearly) laminar regions for their channel flow simulation. By properly designing the transition function, good agreement to temporal DNS results was attained. Voke and Yang (1995) employed the fixed-coefficient Smagorinsky model in conjunction with a low-Reynolds-number correction to simulate bypass transition. Piomelli et al. (1991) studied the energy budget including the SGS terms from DNS data of transitional and turbulent channel flow. They concluded that for an appropriate modelling of both transitional and turbulent channel flow backscatter effects (i.e. energy transfer from small to larger scales) are important.

The class of dynamic SGS models proposed by Germano et al. (1991) calculate their model coefficient adaptively during the simulation. The computation of the model coefficient was subsequently refined by Lilly (1992). The dynamic Smagorinsky model has been successfully applied to, e.g., temporal transition in channel flow (Germano et al., 1991) and spatial transition in incompressible boundary layers (Huai et al., 1997). Several improved versions of the dynamic model exist, e.g. the Lagrangian dynamic SGS model (Meneveau et al., 1996) in which the evolution of the SGS stresses is tracked in a Lagrangian way. The latter model has also been applied to transitional channel flow with good results.

Ducros et al. (1996) introduced the filtered structure function (FSF) model which is also based on the eddy-viscosity assumption. Using the FSF model, the high-pass filter used for the computation of the structure function decreases the influence of long-wave disturbances in the calculation of the SGS terms. As a consequence, the model influence is reduced in regions of the flow which are mainly dominated by mean strain, e.g. in the vicinity of walls or in laminar regions. The FSF model was successfully applied to weakly compressible spatial transition in boundary layer flow. The formation of Λ-vortices and hairpin vortices could clearly be detected, however, no quantitative comparison to experiments or DNS data was given.

The combination of the dynamic Smagorinsky model with the scale-similarity approach (dynamic mixed model, Zang et al. (1993)) yielded very accurate results for the case of a compressible transitional boundary layer at high Mach number (El-Hady and Zang, 1995).

The variational multiscale (VMS) method (Hughes et al., 2000), providing a scale separation between the large-scale fluctuations and the short-wave disturbances, has been used for the simulation of incompressible bypass transition along a flat plate (Calo, 2004). Reasonable agreement with the corresponding DNS (Jacobs and Durbin, 2001; Brandt et al., 2004) has been attained.

3.2 Recent Progress by our Group

In Schlatter (2005), results obtained using large-eddy simulation of transitional and turbulent incompressible channel flow and homogeneous isotropic turbulence are presented. These simulations have been performed using spectral methods in which numerical errors due to differentiation are small and aliasing errors can be avoided (Canuto et al., 1988). For the transition computations, both the temporal and the spatial simulation approach have been employed (Kleiser and Zang, 1991). Various classical and newly devised subgrid-scale closures have been implemented and evaluated, including the approximate deconvolution model (ADM) (Stolz and Adams, 1999), the relaxation-term model (ADM-RT) (Stolz and Adams, 2003; Schlatter et al., 2004a), and the new class of high-pass filtered (HPF) eddy-viscosity models (Stolz et al., 2004, 2005; Schlatter et al., 2005b). These models are discussed briefly in the following.

In order to facilitate the use of deliberately chosen coarse LES grids, the standard ADM methodology (Stolz and Adams, 1999) was revisited. This was necessary due to the observed destabilising properties of the deconvolution operation on such coarse grids in the wall-normal direction. In Schlatter et al. (2004a), in addition to the original ADM algorithm, new variants have been examined, in particular the SGS model based on a direct relaxation regularisation of the velocities (ADM-RT model) which uses a three-dimensional high-pass filtering of the computational quantities. This model is related to the spectral vanishing viscosity (SVV) approach (Karamanos and Karniadakis, 2000). Schlatter et al. (2004b) explore various procedures for the dynamic determination of the relaxation parameter. The appropriate definition of the relaxation term causes the model contributions to vanish during the initial stage of transition and, approximately, in the viscous sublayer of wall turbulence.

The application of the HPF models to transitional channel flow was presented in Stolz et al. (2004, 2005). These models have been proposed independently by Vreman (2003) and Stolz et al. (2004) and are related to the variational multiscale method (Hughes et al., 2000). Detailed analysis of the energy budget including the SGS terms revealed that the contribution to the mean SGS dissipation is nearly zero for the HPF models, while it is a significant part of the SGS dissipation for other SGS models (Schlatter et al., 2005b). Moreover, unlike the classical eddy-viscosity models, the HPF eddy-viscosity models are able to predict backscatter. It has been shown that in channel flow locations with intense backscatter are closely related to low-speed turbulent streaks in both LES and filtered DNS data. In Schlatter et al. (2005b), on the basis of spectral a discretisation a close relationship between the HPF modelling approach and the relaxation term of ADM and ADM-RT could be established. By an accordingly modified high-pass filter, these two approaches become analytically equivalent for homogeneous Fourier directions and constant model coefficients.

The new high-pass filtered (HPF) eddy-viscosity models have also been applied successfully to incompressible forced homogeneous isotropic turbulence with microscale Reynolds numbers Re_λ up to 5500 and to fully turbulent channel flow at moderate Reynolds numbers up to $Re_\tau \approx 590$ (Schlatter et al., 2005b).

Most of the above references show that, e.g. for the model problem of temporal transition in channel flow, spatially averaged integral flow quantities like the skin-friction Reynolds number Re_τ or the shape factor H_{12} of the mean velocity profile can be predicted reasonably well by LES even on comparably coarse meshes, see e.g. Germano et al. (1991); Schlatter et al. (2004a). However, for a reliable LES it is equally important to faithfully represent the physically dominant transitional flow mechanisms and the corresponding three-dimensional vortical structures such as the formation of Λ-vortices and hairpin vortices. A successful SGS model needs to predict those structures well even at low numerical resolution, as demonstrated by Schlatter et al. (2005d, 2006); Schlatter (2005).

The different SGS models have been tested in both the temporal and the spatial transition simulation approach (see Schlatter et al. (2006)). For the spatial simulations, the fringe method has been used to obtain non-periodic flow solutions in the spatially evolving streamwise direction while employing periodic spectral discretisation (Nordström et al., 1999; Schlatter et al., 2005a). The combined effect of the fringe forcing and the SGS model has also been examined. Conclusions derived from temporal results transfer readily to the spatial simulation method, which is more physically realistic but much more computationally expensive.

The computer codes used for the above mentioned simulations have all been parallelised explicitly based on the shared-memory (OpenMP) approach. The codes have been optimised for modern vector and (super-)scalar computer architectures, running very efficiently on different machines from desktop Linux PCs to the NEC SX-5 supercomputer.

4 Conclusions

The results obtained for the canonical case of incompressible channel-flow transition using the various SGS models show that it is possible to accurately simulate transition using LES on relatively coarse grids. In particular, the ADM-RT model, the dynamic Smagorinsky model, the filtered structure-function model and the different HPF models are able to predict the laminar-turbulent changeover. However, the performance of the various models examined concerning an accurate prediction of e.g. the transition location and the characteristic transitional flow structures is considerably different.

By examining instantaneous flow fields from LES of channel flow transition, additional distinct differences between the SGS models can be established. The dynamic Smagorinsky model fails to correctly predict the first stages of breakdown involving the formation of typical hairpin vortices on the coarse LES grid. The no-model calculation, as expected, is generally too noisy during the turbulent breakdown, preventing the identification of transitional structures. In the case of spatial transition, the underresolution of the no-model calculation affects the whole computational domain by producing noisy velocity fluctuations even in laminar flow regions. On the other hand, the ADM-RT model, whose model contributions are confined to the smallest spatial scales, allows for an accurate and physically realistic prediction of the transitional structures even up to later stages of transition. Clear predictions of the one- to the four-spike stages of transition could be obtained. Moreover, the visualisation of the vortical structures shows the appearance of hairpin vortices connected with those stages.

The HPF eddy-viscosity models provide an easy way to implement an alternative to classical fixed-coefficient eddy-viscosity models. The HPF models have shown to perform significantly better than their classical counterparts in the context of wall-bounded shear flows, mainly due to a more accurate description of the near-wall region. The results have shown that a fixed model coefficient is sufficient for the flow cases considered. No dynamic procedure for the determination of the model coefficient was found necessary, and no empirical wall-damping functions were needed.

To conclude, LES using advanced SGS models are able to faithfully simulate flows which contain intermittent laminar, turbulent and transitional regions.

References

J. Bardina, J. H. Ferziger, and W. C. Reynolds. Improved subgrid models for large-eddy simulation. *AIAA Paper*, 1980-1357, 1980.

L. Brandt, P. Schlatter, and D. S. Henningson. Transition in boundary layers subject to free-stream turbulence. *J. Fluid Mech.*, 517:167–198, 2004.

V. M. Calo. *Residual-based multiscale turbulence modeling: Finite volume simulations of bypass transition.* PhD thesis, Stanford University, USA, 2004.

C. Canuto, M. Y. Hussaini, A. Quarteroni, and T. A. Zang. *Spectral Methods in Fluid Dynamics.* Springer, Berlin, Germany, 1988.

J. A. Domaradzki and N. A. Adams. Direct modelling of subgrid scales of turbulence in large eddy simulations. *J. Turbulence*, 3, 2002.

F. Ducros, P. Comte, and M. Lesieur. Large-eddy simulation of transition to turbulence in a boundary layer developing spatially over a flat plate. *J. Fluid Mech.*, 326:1–36, 1996.

N. M. El-Hady and T. A. Zang. Large-eddy simulation of nonlinear evolution and breakdown to turbulence in high-speed boundary layers. *Theoret. Comput. Fluid Dynamics*, 7:217–240, 1995.

M. Germano, U. Piomelli, P. Moin, and W. H. Cabot. A dynamic subgrid-scale eddy viscosity model. *Phys. Fluids A*, 3(7):1760–1765, 1991.

B. J. Geurts. *Elements of Direct and Large-Eddy Simulation.* Edwards, Philadelphia, USA, 2004.

N. Gilbert and L. Kleiser. Near-wall phenomena in transition to turbulence. In S. J. Kline and N. H. Afgan, editors, *Near-Wall Turbulence – 1988 Zoran Zarić Memorial Conference*, pages 7–27. Hemisphere, New York, USA, 1990.

X. Huai, R. D. Joslin, and U. Piomelli. Large-eddy simulation of transition to turbulence in boundary layers. *Theoret. Comput. Fluid Dynamics*, 9:149–163, 1997.

T. J. R. Hughes, L. Mazzei, and K. E. Jansen. Large eddy simulation and the variational multiscale method. *Comput. Visual. Sci.*, 3:47–59, 2000.

R. G. Jacobs and P. A. Durbin. Simulations of bypass transition. *J. Fluid Mech.*, 428: 185–212, 2001.

J. Jeong and F. Hussain. On the identification of a vortex. *J. Fluid Mech.*, 285:69–94, 1995.

Y. S. Kachanov. Physical mechanisms of laminar-boundary-layer transition. *Annu. Rev. Fluid Mech.*, 26:411–482, 1994.

G.-S. Karamanos and G. E. Karniadakis. A spectral vanishing viscosity method for large-eddy simulations. *J. Comput. Phys.*, 163:22–50, 2000.

L. Kleiser and T. A. Zang. Numerical simulation of transition in wall-bounded shear flows. *Annu. Rev. Fluid Mech.*, 23:495–537, 1991.

M. Lesieur and O. Métais. New trends in large-eddy simulations of turbulence. *Annu. Rev. Fluid Mech.*, 28:45–82, 1996.

D. K. Lilly. A proposed modification of the Germano subgrid-scale closure method. *Phys. Fluids A*, 4(3):633–635, 1992.

C. Meneveau and J. Katz. Scale-invariance and turbulence models for large-eddy simulation. *Annu. Rev. Fluid Mech.*, 32:1–32, 2000.

C. Meneveau, T. S. Lund, and W. H. Cabot. A Lagrangian dynamic subgrid-scale model of turbulence. *J. Fluid Mech.*, 319:353–385, 1996.

P. Moin and K. Mahesh. Direct numerical simulation: A tool in turbulence research. *Annu. Rev. Fluid Mech.*, 30:539–578, 1998.

J. Nordström, N. Nordin, and D. S. Henningson. The fringe region technique and the Fourier method used in the direct numerical simulation of spatially evolving viscous flows. *SIAM J. Sci. Comput.*, 20(4):1365–1393, 1999.

U. Piomelli. Large-eddy and direct simulation of turbulent flows. In *CFD2001 – 9e conférence annuelle de la société Canadienne de CFD.* Kitchener, Ontario, Canada, 2001.

U. Piomelli, W. H. Cabot, P. Moin, and S. Lee. Subgrid-scale backscatter in turbulent and transitional flows. *Phys. Fluids A*, 3(7):1799–1771, 1991.

U. Piomelli, T. A. Zang, C. G. Speziale, and M. Y. Hussaini. On the large-eddy simulation of transitional wall-bounded flows. *Phys. Fluids A*, 2(2):257–265, 1990.

D. Rempfer. Low-dimensional modeling and numerical simulation of transition in simple shear flows. *Annu. Rev. Fluid Mech.*, 35:229–265, 2003.

P. Sagaut. *Large Eddy Simulation for Incompressible Flows.* Springer, Berlin, Germany, 3rd edition, 2005.

P. Schlatter. *Large-eddy simulation of transition and turbulence in wall-bounded shear flow.* PhD thesis, ETH Zürich, Switzerland, Diss. ETH No. 16000, 2005. Available online from `http://e-collection.ethbib.ethz.ch`.

P. Schlatter, N. A. Adams, and L. Kleiser. A windowing method for periodic inflow/outflow boundary treatment of non-periodic flows. *J. Comput. Phys.*, 206(2): 505–535, 2005a.

P. Schlatter, S. Stolz, and L. Kleiser. LES of transitional flows using the approximate deconvolution model. *Int. J. Heat Fluid Flow*, 25(3):549–558, 2004a.

P. Schlatter, S. Stolz, and L. Kleiser. Relaxation-term models for LES of transitional/turbulent flows and the effect of aliasing errors. In R. Friedrich, B. J. Geurts, and O. Métais, editors, *Direct and Large-Eddy Simulation V*, pages 65–72. Kluwer, Dordrecht, The Netherlands, 2004b.

P. Schlatter, S. Stolz, and L. Kleiser. Evaluation of high-pass filtered eddy-viscosity models for large-eddy simulation of turbulent flows. *J. Turbulence*, 6(5), 2005b.

P. Schlatter, S. Stolz, and L. Kleiser. LES of spatial transition in plane channel flow. *J. Turbulence*, 2006. To appear.

P. Schlatter, S. Stolz, and L. Kleiser. Applicability of LES models for prediction of transitional flow structures. In R. Govindarajan, editor, *Laminar-Turbulent Transition.* Sixth IUTAM Symposium 2004 (Bangalore, India), Springer, Berlin, Germany, 2005d.

P. J. Schmid and D. S. Henningson. *Stability and Transition in Shear Flows.* Springer, Berlin, Germany, 2001.

J. Smagorinsky. General circulation experiments with the primitive equations. *Mon. Weath. Rev.*, 91(3):99–164, 1963.

S. Stolz and N. A. Adams. An approximate deconvolution procedure for large-eddy simulation. *Phys. Fluids*, 11(7):1699–1701, 1999.

S. Stolz and N. A. Adams. Large-eddy simulation of high-Reynolds-number supersonic boundary layers using the approximate deconvolution model and a rescaling and recycling technique. *Phys. Fluids*, 15(8):2398–2412, 2003.

S. Stolz, N. A. Adams, and L. Kleiser. An approximate deconvolution model for large-eddy simulation with application to incompressible wall-bounded flows. *Phys. Fluids*, 13(4):997–1015, 2001a.

S. Stolz, N. A. Adams, and L. Kleiser. The approximate deconvolution model for large-eddy simulations of compressible flows and its application to shock-turbulent-boundary-layer interaction. *Phys. Fluids*, 13(10):2985–3001, 2001b.

S. Stolz, P. Schlatter, and L. Kleiser. High-pass filtered eddy-viscosity models for large-eddy simulations of transitional and turbulent flow. *Phys. Fluids*, 17:065103, 2005.

S. Stolz, P. Schlatter, D. Meyer, and L. Kleiser. High-pass filtered eddy-viscosity models for LES. In R. Friedrich, B. J. Geurts, and O. Métais, editors, *Direct and Large-Eddy Simulation V*, pages 81–88. Kluwer, Dordrecht, The Netherlands, 2004.

E. R. van Driest. On the turbulent flow near a wall. *J. Aero. Sci.*, 23:1007–1011, 1956.

P. Voke and Z. Yang. Numerical study of bypass transition. *Phys. Fluids*, 7(9):2256–2264, 1995.

A. W. Vreman. The filtering analog of the variational multiscale method in large-eddy simulation. *Phys. Fluids*, 15(8):L61–L64, 2003.

Y. Zang, R. L. Street, and J. R. Koseff. A dynamic mixed subgrid-scale model and its application to turbulent recirculating flows. *Phys. Fluids A*, 5(12):3186–3196, 1993.

Applications I: Finite Element Method

Computational Efficiency of Parallel Unstructured Finite Element Simulations

Malte Neumann[1], Ulrich Küttler[2], Sunil Reddy Tiyyagura[3], Wolfgang A. Wall[2], and Ekkehard Ramm[1]

[1] Institute of Structural Mechanics, University of Stuttgart, Pfaffenwaldring 7, D-70550 Stuttgart, Germany, {neumann,ramm}@statik.uni-stuttgart.de, WWW home page: http://www.uni-stuttgart.de/ibs/

[2] Chair of Computational Mechanics, Technical University of Munich, Boltzmannstraße 15, D-85747 Garching, Germany, {kuettler,wall}@lnm.mw.tum.de, WWW home page: http://www.lnm.mw.tum.de/

[3] High Performance Computing Center Stuttgart (HLRS), Nobelstraße 19, D-70569 Stuttgart, Germany, sunil@hlrs.de, WWW home page: http://www.hlrs.de/

Abstract In this paper we address various efficiency aspects of finite element (FE) simulations on vector computers. Especially for the numerical simulation of large scale Computational Fluid Dynamics (CFD) and Fluid-Structure Interaction (FSI) problems efficiency and robustness of the algorithms are two key requirements.

In the first part of this paper a straightforward concept is described to increase the performance of the integration of finite elements in arbitrary, unstructured meshes by allowing for vectorization. In addition the effect of different programming languages and different array management techniques on the performance will be investigated.

Besides the element calculation, the solution of the linear system of equations takes a considerable part of computation time. Using the jagged diagonal format (JAD) for the sparse matrix, the average vector length can be increased. Block oriented computation schemes lead to considerably less indirect addressing and at the same time packaging more instructions. Thus, the overall performance of the iterative solver can be improved.

The last part discusses the input and output facility of parallel scientific software. Next to efficiency the crucial requirements for the IO subsystem in a parallel setting are scalability, flexibility and long term reliability.

1 Introduction

The ever increasing computation power of modern computers enable scientists and engineers alike to approach problems that were unfeasible only years ago. There are, however, many kinds of problems that demand computation power only highly parallel clusters or advanced supercomputers are able to provide. Various of these, like multi-physics and multi-field problems (e.g. the interaction of fluids and structures), play an important role for both their engineering relevance and scientific challenges. This amounts to the need for highly parallel computation facilities, together with specialized software that utilizes these parallel machines.

The work described in this paper was done on the basis of the research finite element program CCARAT, that is jointly developed and maintained at the Institute of Structural Mechanics of the University of Stuttgart and the Chair of Computational Mechanics at the Technical University of Munich. The research code CCARAT is a multipurpose finite element program covering a wide range of applications in computational mechanics, like e.g. multi-field and multi-scale problems, structural and fluid dynamics, shape and topology optimization, material modeling and finite element technology. The code is parallelized using MPI and runs on a variety of platforms, on single processor systems as well as on clusters.

After a general introduction on computational efficiency and vector processors three performance aspects of finite elements simulations are addressed: In the second chapter of this paper a straightforward concept is described to increase the performance of the integration of finite elements in arbitrary, unstructured meshes by allowing for vectorization. The following chapter discusses the effect of different matrix storage formats on the performance of an iterative solver and last part covers the input and output facility of parallel scientific software. Next to efficiency the crucial requirements for the IO subsystem in a parallel setting are scalability, flexibility and long term reliability.

1.1 Computational Efficiency

For a lot of todays scientific applications, e.g. the numerical simulation of large scale Computational Fluid Dynamics (CFD) and Fluid-Structure Interaction (FSI) problems, computing time is still a limiting factor for the size and complexity of the problem, so the available computational resources must be used most efficiently. This especially concerns superscalar processors where the gap between sustained and peak performance is growing for scientific applications. Very often the sustained performance is below 5 percent of peak. The efficiency on vector computers is usually much higher. For vectorizable programs it is possible to achieve a sustained performance of 30 to 60 percent, or above of the peak performance [1, 2].

Starting with a low level of serial efficiency, e.g. on a superscalar computer, it is a reasonable assumption that the overall level of efficiency of the code will drop even further when run in parallel. Therefore looking at the serial efficiency is one key ingredient for a highly efficient parallel code [1].

To achieve a high efficiency on a specific system it is in general advantageous to write hardware specific code, i.e. the code has to make use of the system specific features like vector registers or the cache hierarchy. As our main target architectures are the NEC SX-6+ and SX-8 parallel vector computers, we will address some aspects of vector optimization in this paper. But as we will show later this kind of performance optimization also has a positive effect on the performance of the code on other architectures.

1.2 Vector Processors

Vector processors like the NEC SX-6+ or SX-8 processors use a very different architectural approach than conventional scalar processors. Vectorization exploits regularities in the computational structure to accelerate uniform operations on independent data sets. Vector arithmetic instructions involve identical operations on the elements of vector operands located in the vector registers. A lot of scientific codes like FE programs allow vectorization, since they are characterized by predictable fine-grain data-parallelism [2].

For non-vectorizable instructions the SX machines also contain a cache-based superscalar unit. Since the vector unit is significantly more powerful than this scalar processor, it is critical to achieve high vector operations ratios, either via compiler discovery or explicitly through code and data (re-)organization.

In recognition of the opportunities in the area of vector computing, the High Performance Computing Center Stuttgart (HLRS) and NEC are jointly working on a cooperation project "Teraflop Workbench", which main goal is to achieve sustained teraflop performance for a wide range of scientific and industrial applications. The hardware platforms available in this project are:

NEC SX-8: 72 nodes, 8 CPUs per node, 16 Gflops vector peak performance per CPU (2 GHz clock frequency), Main memory bandwidth of 64 GB/s per CPU, Internode bandwidth of 16 GB/s per node

NEC SX-6+: 6 nodes, 8 CPUs per node, 9 Gflops vector peak performance per CPU (0.5625 GHz clock frequency), Main memory bandwidth of 36 GB/s per CPU, Internode bandwidth of 8 GB/s per node

NEC TX7: 32 Itanium2 CPUs, 6 Gflops peak performance per CPU

NEC Linux Cluster: 200 nodes, 2 Intel Nocona CPUs per node, 6.4 Gflops peak performance per CPU, Internode bandwidth of 1 GB/s

An additional goal is to establish a complete pre-processing – simulation – post-processing – visualization workflow in an integrated and efficient way using the above hardware resources.

1.3 Vector Optimization

To achieve high performance on a vector architecture there are three main variants of vectorization tuning:

- compiler flags
- compiler directives
- code modifications.

The usage of *compiler flags* or *compiler directives* is the easiest way to influence the vector performance, but both these techniques rely on the existence of vectorizable code and on the ability of the compiler to recognize it. Usually the resulting performance will not be as good as desired.

In most cases an optimal performance on a vector architecture can only be achieved with code that was especially designed for this kind of processor. Here the data management as well as the structure of the algorithms are important. But often it is also very effective for an existing code to concentrate the vectorization efforts on performance critical parts and use more or less extensive *code modifications* to achieve a better performance. The reordering or fusion of loops to increase the vector length or the usage of temporary variables to break data dependencies in loops can be simple measures to improve the vector performance.

2 Vectorization of Finite Element Integration

For the numerical solution of large scale CFD and FSI problems usually highly complex, stabilized elements on unstructured grids are used. The element evaluation and assembly for these elements is often, besides the solution of the system of linear equations, a main time consuming part of a finite element calculation. Whereas a lot of research is done in the area of solvers and their efficient implementation, there is hardly any literature on efficient implementation of advanced finite element formulations. Still a large amount of computing time can be saved by an expert implementation of the element routines. We would like to propose a straightforward concept, that requires only little changes to an existing FE code, to improve significantly the performance of the integration of element matrices of an arbitrary unstructured finite element mesh on vector computers.

2.1 Sets of Elements

The main idea of this concept is to group computationally similar elements into sets and then perform all calculations necessary to build the element matrices simultaneously for all elements in one set. *Computationally similar* in this context means, that all elements in one set require exactly the same operations to integrate the element matrix, that is each set consists of elements with the same topology and the same number of nodes and integration points.

The changes necessary to implement this concept are visualized in the structure charts in Fig. 1. Instead of looping all elements and calculating the element matrix individually, now all sets of elements are processed. For every set the usual procedure to integrate the matrices is carried out, except on the lowest level, i.e. as the innermost loop, a new loop over all elements in the current set is introduced. This loop suits especially vector machines perfectly, as the calculations inside are quite simple and, most important, consecutive steps do not depend on each other. In addition the length of this loop, i.e. the size of the element sets, can be chosen freely, to fill the processor's vector pipes.

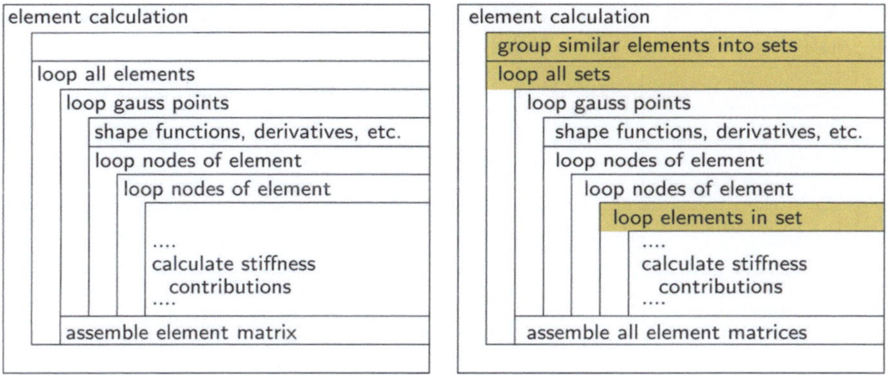

Fig. 1. Old (left) and new (right) structure of an algorithm to evaluate element matrices

The only limitation for the size of the sets are additional memory requirements, as now intermediate results have to be stored for all elements in one set. For a detailed description of the dependency of the size of the sets and the processor type see Sect. 2.2.

2.2 Further Influences on the Efficiency

Programming Language & Array Management

It is well known that the programming language can have a large impact on the performance of a scientific code. Despite considerable effort on other languages [3, 4] Fortran is still considered the best choice for highly efficient code [5] whereas some features of modern programming languages, like pointers in C or objects in C++, make vectorization more complicated or even impossible [2].

Especially the very general pointer concept in C makes it difficult for the compiler to identify data-parallel loops, as different pointers might alias each other. There are a few remedies for this problem like compiler flags or the *restrict* keyword. The latter is quite new in the C standard and it seems that it is not yet fully implemented in every compiler.

We have implemented the proposed concept for the calculation of the element matrices in 5 different variants. The first four of them are implemented in C, the last one in Fortran. Further differences are the array management and the use of the restrict keyword. For a detailed description of the variants see Table 1. Multi-dimensional arrays denote the use of 3- or 4-dimensional arrays to store intermediate results, whereas one-dimensional arrays imply a manual indexing.

The results in Table 1 give the cpu time spent for the calculation of some representative element matrix contributions standardized by the time used by the original code. The positive effect of the grouping of elements can be clearly seen for the vector processor. The calculation time is reduced to less than 3% for all variants. On the other two processors the grouping of elements does not result

Table 1. Influences on the performance. Properties of the five different variants and their relative time for calculation of stiffness contributions

	orig	var1	var2	var3	var4	var5
language array dimensions restrict keyword	C multi	C multi	C multi restrict	C one	C one restrict	Fortran multi
SX-6+[1]	1.000	0.024	0.024	0.016	0.013	0.011
Itanium2[2]	1.000	1.495	1.236	0.742	0.207	0.105
Pentium4[3]	1.000	2.289	1.606	1.272	1.563	0.523

in a better performance for all cases. The Itanium architecture shows only an improved performance for one dimensional array management and the variant implemented in Fortran and the Pentium processor performs in general worse for the new structure of the code. Only for the last variant the calculation time is cut in half.

It can be clearly seen, that the effect of the restrict keyword varies for the different compilers/processors and also for one-dimensional and multi-dimensional arrays. Using restrict on the SX-6+ results only in small improvements for one-dimensional arrays, on the Itanium architecture the speed-up for this array management is even considerable. In contrast to this on the Pentium architecture the restrict keyword has a positive effect on the performance of multi-dimensional arrays and a negative effect for one-dimensional ones.

The most important result of this analysis is the superior performance of Fortran. This is the reason we favor Fortran for performance critical scientific code and use the last variant for our further examples.

Size of the Element Sets

As already mentioned before the size of the element sets and with it the length of the innermost loop needs to be different on different hardware architectures. To find the optimal sizes on the three tested platforms we measured the time spent in one subroutine, which calculates representative element matrix contributions, for different sizes of the element sets (Fig. 2).

For the cache based Pentium4 processor the best performance is achieved for very small sizes of the element sets. This is due to the limited size of cache which usage is crucial for performance. The best performance for the measured subroutine was achieved with 12 elements per set.

---

[1] NEC SX-6+, 565 MHz; NEC C++/SX Compiler, Version 1.0 Rev. 063; NEC FORTRAN/SX Compiler, Version 2.0 Rev. 305.

[2] Hewlett Packard Itanium2, 1.3 GHz; HP aC++/ANSI C Compiler, Rev. C.05.50; HP F90 Compiler, v2.7.

[3] Intel Pentium4, 2.6 GHz; Intel C++ Compiler, Version 8.0; Intel Fortran Compiler, Version 8.0.

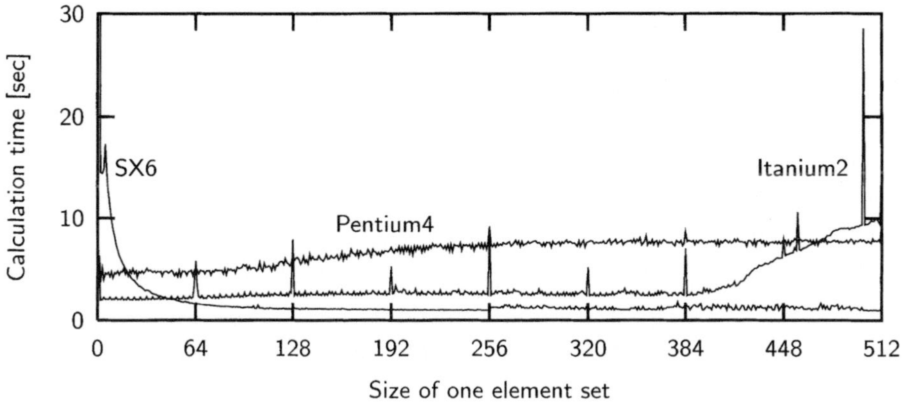

Fig. 2. Calculation time for one subroutine that calculates representative element matrix contributions for different sizes of one element set

The Itanium2 architecture shows an almost constant performance for a large range of sizes. The best performance is achieved for a set size of 23 elements.

For the vector processor SX-6+ the calculation time decrease for growing sizes up to 256 elements per set, which corresponds to the size of the vector registers. For larger sets the performance only varies slightly with optimal values for multiples of 256.

2.3 Results

Concluding we would like to demonstrate the positive effect of the proposed concept for the calculation of element matrices on a full CFD simulation. The flow is the Beltrami-Flow (for details see [6]) and the unit-cube was discretized by 32768 stabilized 8-noded hexahedral elements [7].

In Fig. 3 the total calculation time for 32 time steps of this example and the fractions for the element calculation and the solver on the SX-6+ are given for the original code and the full implementation of variant 5. The time spent for the element calculation, formerly the major part of the total time, could be reduced by a factor of 24.

This considerable improvement can also be seen in the sustained performance given in Table 2 as percentage of peak performance. The original code not written for any specific architecture has only a poor performance on the SX-6+ and a moderate one on the other platforms. The new code, designed for a vector processor, achieves for the complete element calculation an acceptable efficiency of around 30% and for several subroutines, like the calculation of some stiffness contributions, even a superior efficiency of above 70%. It has to be noted that these high performance values come along with a vector length of almost 256 and a vector operations ratio of above 99.5%.

But also for the Itanium2 and Pentium4 processors, which were not the main target architectures, the performance was improved significantly and for

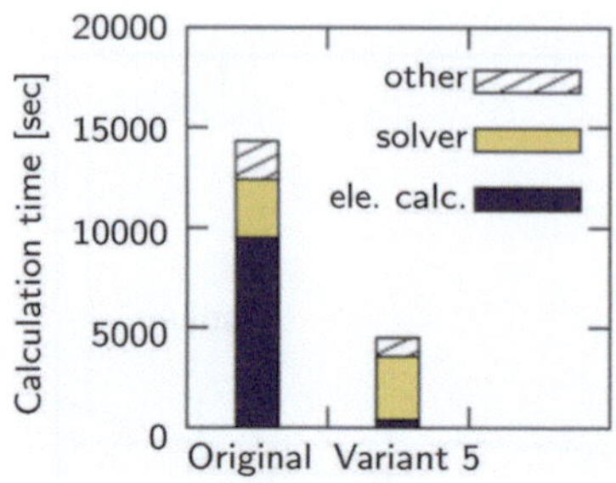

Fig. 3. Split-up of total calculation time for 32 time steps of the Beltrami Flow on the SX-6+

	element calc.		stiffness contr.	
	original	var5	original	var5
SX-6+	0.95	29.55	0.83	71.07
Itanium2	8.68	35.01	6.59	59.71
Pentium4	12.52	20.16	10.31	23.98

Table 2. Efficiency of original and new code in percent of peak performance

the Itanium2 the new code reaches around the same efficiency as on the vector architecture.

3 Iterative Solvers

CCARAT uses external solvers such as Aztec to solve the linear system of equations. Most of the public domain iterative solvers are optimized for performance only on cache based machines, hence they do not performance well on vector systems. The main reason for this is the storage formats used in these packages, which are mostly row or column oriented.

The present effort is directed at improving the efficiency of the iterative solvers on vector machines. The most important kernel operation of any iterative solver is the matrix vector multiplication. We shall look at the efficiency of this operation, especially on vector architectures, where its performance is mainly affected by the average vector length and the frequency of indirect addressing.

3.1 Sparse Storage Formats

Short vector length is a classical problem that affects the performance on vector systems. The reason for short vector lengths in this case is the sparse storage format used. Most of the sparse linear algebra libraries implement either a row oriented or a column oriented storage format. In these formats, the non-zero entries in each row or a column are stored successively. This number usually turns out to be smaller than the effective size of the vector pipes on SX (which is 256 on SX-6+ and SX-8). Hence, both these formats lead to short vector lengths at runtime. The only way to avoid this problem is to use a pseudo diagonal format. This format ensures that, at least the length of the first few non-zero pseudo diagonals is equivalent to the size of the matrix. Hence, it overcomes the problem of short vector length. An example of such a format is the well known jagged diagonal format (JAD). The performance data with row and diagonal formats on SX-6+ and SX-8 is listed in Table 3.

Table 3. Performance (per CPU) of row and diagonal formats on SX-6+/SX-8

Machine	Format	MFlops	Bank conflicts (%)
SX-6+	Row storage	510	2.8
SX-6+	Diagonal storage	1090	2.2
SX-8	Row storage	770	4.3
SX-8	Diagonal storage	1750	1.3

It is clear from the data stated in Table 3 that diagonal formats are at least twice as more efficient as row or column formats. The superiority in performance is simply because of better vector lengths. The following is a skeleton of a sparse matrix vector multiplication algorithm:

```
for var = 0, rows/cols/diags
    offset = index(val)
    for len = 0, row/col/diag length
      res(val/len) += mat(offset+len) * vec(index(offset+len))
    end for
end for
```

Figure 4 shows the timing diagram, where the execution of vector operations and their performance can be estimated. The gap between the measured and the peak performance can also be easily understood with the help of this figure. The working of the load/store unit along with both the functional units is illustrated here. The load/store unit can do a single pipelined vector load or a vector store at a time, which takes 256 (vector length on SX) CPU cycles. Each functional unit can perform a pipelined floating point vector operation at a time, each of which takes 256 CPU cycles. It is to be noted that the order of the actual load/store and FP instructions can be different from the one shown in this figure. But, the effective number of vector cycles needed remain the same.

From Fig. 4, it can be inferred that most of the time for computation is spent in loading and storing the data. There are only two effective floating point vector operations in 5 vector cycles (10 possible FP operations). So, the expected performance from this operation is 2/10 of the peak (16 Gflops per CPU on SX-8). But indirect addressing of the vector further affects this expected performance

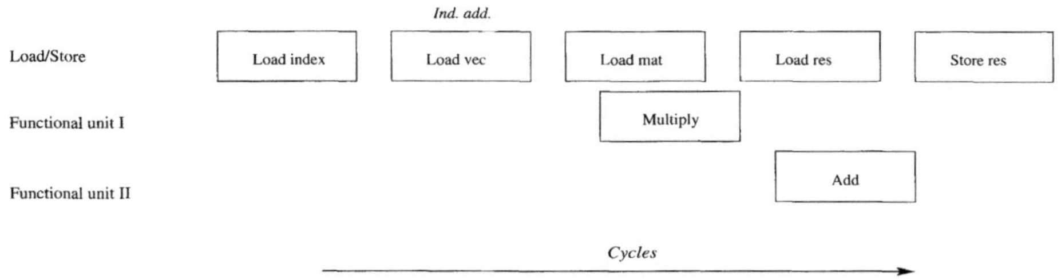

Fig. 4. Timing diagram for sparse matrix-vector multiplication

(3.2 Gflops per CPU on SX-8), resulting in 1.75 Gflops per CPU. This can be slightly improved by avoiding the unnecessary loading of the result vector (strip mining the inner loop). To enable this, a vector register (size equivalent to the vector pipeline length) has to be allocated and the temporary results are stored in it. Then the results are copied back to the result vector at the end of each stripped loop. This will save loading the result vector in each cycle as shown in Fig. 4, thereby improving the performance. Similar techniques are also adapted for gaining performance on other vector architectures, like the CRAY X1[8]. With vector register allocation, a performance improvement of around 25% was noticed for matrix vector multiplication using diagonal format. It is worthwhile to note that, on SX, this feature can only be used with fortran and not yet with C/C++. Making sure that the vector pipelines are filled (with the pseudo diagonal storage structure) still only doubles the performance. In most of the cases, this problem is relatively simple to overcome.

3.2 Indirect Addressing

Indirect addressing also poses a great threat to performance. The overhead depends not only on the efficiency of the hardware implementation to handle it, but also on the amount of memory bank conflicts it creates (problem dependent). For the matrix vector multiplication, loading an indirectly addressed vector took 3–4 times longer than to load a directly addressed one. So, this gives a rough estimation of the extent of the problem created by indirect addressing. The actual effect in the sense of floating point performance has to be doubled as both the functional units would be idle during these cycles.

The theoretical peak can only be achieved if both the functional units operate at the same time. If a simple computation never uses both the functional units at the same time, then the theoretically attainable peak is reduced to half. The next question is how to keep both the functional units working and also to reduce the amount of indirect addressing required. Operating on small blocks looks to be a promising solution.

3.3 Block Computations

The idea of block computations originates from the fact that many problems have multiple physical variables per node. So, small blocks can be formed by grouping the equations at each node. This has a tremendous effect on performance. There are mainly two reasons behind this enormous improvement in performance. Firstly, it reduces the amount of indirect addressing required. Secondly, both the functional units are used at the same time (at least in some cycles). The reduction in indirect addressing can be seen from the following block matrix vector multiplication algorithm:

```
for var = 0, rows/cols/diags of blocks(3x3)
  offset = index(var)
  for len = 0, row/col/diag length
```

```
    res(var) += mat(offset+len)   * vec(index(offset+len))
              + mat(offset+len+1) * vec(index(offset+len)+1)
              + mat(offset+len+2) * vec(index(offset+len)+2)
    res(var+1) += ...// 'vec' is reused
    res(var+2) += ...// 'vec' is reused
  end for
end for
```

So, for each matrix block, the vector block to be multiplied is indirectly addressed only thrice. These vector quantities are then reused. On the whole, indirect addressing is reduced by a factor equivalent to block size. This along with the improved use of functional units (illustrated in figure 5 for 3×3 blocks), results in an improved performance. The expected performance for directly addressed vectors is around 9.6 Gflops per CPU (18 FP operations in 15 vector cycles) for 3×3 blocks. But, including the overhead due to indirect addressing, the resulting performance is around 6.0 Gflops per CPU.

This is an elegant way to achieve a good portion of the theoretical peak performance on the vector machine. Block operations are not only efficient on vector systems, but also on scalar architectures [9]. The results of matrix vector multiplication with blocks are included in Table 4. The block size 4+ means that an empty extra element is allocated after each matrix block to avoid bank conflicts (due to even strides). This is the only disadvantage of working with blocks. Anyway, it can be overcome with simple techniques such as array padding. One can also notice the improvement in performance for even strides by comparing the performance for 4×4 blocks on SX-6+ and SX-8. This happens to be more than the theoretical factor of 1.78. So, a part of it is due to the improved hard-

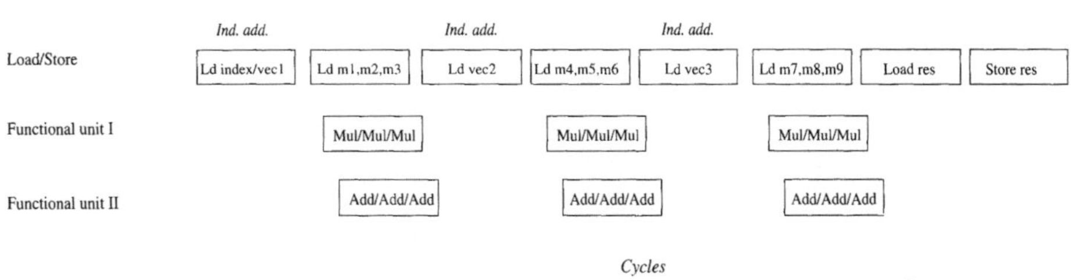

Fig. 5. Timing diagram for block sparse matrix vector multiplication

Table 4. Performance of diagonal block format on SX-6+ and SX-8

Machine	Block size	MFlops	Bank conflicts (%)
SX-6+	4	675	85.8
SX-6+	4+	3410	4.6
SX-6+	5	2970	10.6
SX-8	4	1980	76.1
SX-8	4+	5970	9.7
SX-8	5	5800	5.9

ware implementation for even strides on SX-8. Operating on blocks has also the advantage of using block preconditioning techniques, which are considered to be numerically superior and at the same time perform well on vector machines [10].

4 Parallel Input and Output

Most of the time the input and output facility of scientific software only gets little attention. For moderate scale simulations the handling of input and output is purely guided by convenience considerations, which is why many scientific software systems deploy only very simple IO facilities. On modern supercomputers, all of which are highly parallel, more considerations come into play. The input and output subsystem must take advantage of the parallel environment in order to achieve sufficient execution speed. Other requirements like long term reliability are of increasing importance, too. And still usability and convenience do stay important issues, because people that work on huge scale simulations face difficulties enough without worrying about IO subtleties.

This section describes design and implementation of the parallel IO subsystem of CCARAT. The IO subsystem was specifically designed to enable CCARAT to take advantage of highly parallel computer systems, thus execution speed and scalability are prominent among its design goals.

4.1 Requirements for IO Subsystems in a Parallel Setting

A usual finite element simulation of FSI problems consists of one input call followed by a possibly large number of time step calculations. At the end of each time step calculation results are written. So the more critical IO operation, from a performance point of view, is output.

The results a Finite Element code needs to write in each time step have a comparatively simple structure. There are nodal results like displacements, velocities or pressure. This kind of results are very uniform, there is one scalar or vector entry for each node of the mesh. Other results might be less simple structured, for example things like stress, stress history or crack information could be relevant, depending very much on the problem type under consideration. This kind of results is usually attached to elements and as there is no limit to the physical content of the elements a very wide range of element specific result types are possible. In general it is not feasible to predict all possible element output types. This demands a certain degree of flexibility in the output system, yet the structures inside an element are hardly very complex from the data handling point of view. A third kind of output to be written are restart information. To restart a calculation internal state variables must be restored, but since one and the same program reads and writes these variables these output operations come down to a simple memory dump. That is there are no complex hierarchical structures involved in any of the result types. The challenge in all three of them is the amount of result data the code must handle.

Together with the above discussion these considerations lead to four major requirements for the IO subsystem. In a nutshell these are as follows:

Simplicity Nobody appreciates unnecessary complexities. The simple structure of the result data should be reflected in the design of the IO subsystem.

Efficiency On parallel computers the output, too, must be done in parallel in order to scale.

Flexibility The output system must work with a wide range of algorithms including future algorithms that have not been invented yet. It must work, too, with a wide range of hardware platforms, facilitating data exchange between all of them.

Reliability The created files will stay around for many years even though the code continuously develops. So a clear file format is needed that contains all information necessary to interpret the files.

4.2 Design of a Parallel IO Subsystem

These requirements motivate the following design decisions:

One IO Implementation for all Algorithms

Many finite element codes, like CCARAT, can solve a variety of different types of problems. It is very reasonable to introduce just one IO subsystem that serves them all. That is the basic assumption grounding the rest of the discussion.

No External IO Libraries

The need for input and output in scientific computation is as old as scientific computation itself. Consequently there are well established libraries, for example the Hierarchical Data Format [11] or the Network Common Data Form [12], that provide facilities for reading and writing any kind of data on parallel machines. It seemed to us, however, that these libraries provided much more that we desired. The ability to write deeply nested structures, for instant, is not required at all. And at the same time we felt uncomfortable with the library depended file formats that nobody can read other than the library that created the files. The compelling reason not to use any of those libraries, however, was the fear to rely on a library that might not be available for our next machine. The fewer external libraries a code uses the easier it can be ported to new platforms.

Postprocessing by Filter Applications

We anticipate the need to process our results in many ways, using different postprocessing tools that in turn require different file formats. Obviously we cannot write our results using all the different formats we will probably need. Instead we

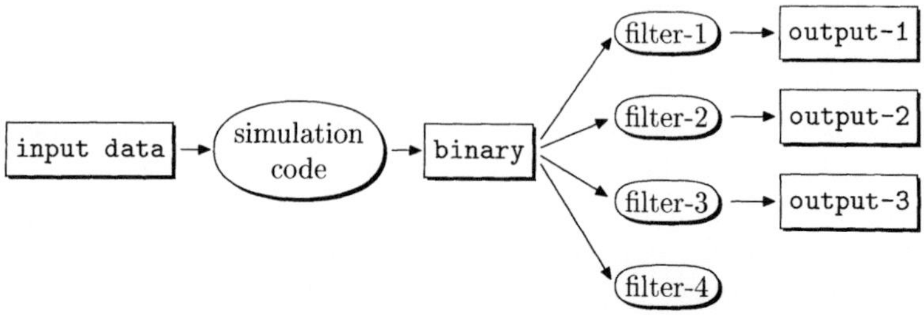

Fig. 6. The call procedure from CCARAT input data to postprocessor specific output files. The rectangles symbolize different file formats, the ellipses represent the programs that read and write these files. Filter programs in this figure are just examples for a variety of postprocessing options

write the results just once and generate special files that our postprocessors demand by external filter programs. There can be any number of filters and we are able to write new ones when a new postprocessor comes along, so we gain great flexibility. Figure 6 depicts the general arrangement with four example filters. However, we have to pay the price of one more layer, one more postprocessing step. This can be costly because of the huge amount of data. But then the benefits are not just that we can use any postprocessor we like, we are also freed from worrying about postprocessors while we write the results. That is we can choose a file format that fits the requirements defined above and do not need to care whether there are applications that want to have the results in that format. It is this decision that enables us to implement a simple, efficient, flexible and reliable IO system.

4.3 File Format for Parallel IO

With above decisions in place the choice of a file format is hardly more than an implementation detail.

Split of Control Information and Binary Data Files

Obviously the bulk data files need to be binary. But we want these files to be as simple as possible, so we write integer and double values to different files. This way we obtain files that contain nothing but one particular type of number, which enables us to easily access the raw numbers with very basic tools, even shell scripts will do if the files are small enough. Of course we do not intend to read the numbers by hand regularly, but it is important to know how to get at the numbers if we have to. Furthermore we decided to create only big-endian files in order be platform independent.

On the other hand we need some kind of control structure to be stored. After all, there might be many time steps each of which contributes a number of results. If we store the results consecutively in the data files we will have to know the places where one result ends and another one starts. For this purpose we introduce a control file. This one will not be large so it is perfectly fine to use plain text, the best format for storing knowledge, see Hunt and Thomas [13].

The interesting point about text files is how to read them back. In the case of the control file we decided in favour of a very simple top down parser that follows closely the one given by Aho, Sethi and Ullman [14]. That is we have a very compact context free grammar definition, consisting of just three rules. On reading a control file we create a syntax tree that can easily be traversed afterwards. That way the control files are easy to read by human beings, containing very little noise compared to xml files for instance, and yet we obtain an easy to use hierarchical structure. That is a crucial feature both for the flexibility we need and for the simplicity that is required.

Node and Element Values Sorted

As said before we have to write values for all nodes or all elements of a mesh. So one particular result consists of entries of constant length, one entry for each node or element. We write these entries consecutively, ordered by node or element id.[4] This arrangement greatly facilitates access to the results.

No Processor Specific Files

On a parallel computer each processor could easily write its own file with the results calculated by this processor and all the output would be perfectly parallel. This, however, puts the burden of gathering and merging the result files on the user and the postprocessor has to achieve what the solver happily avoided. There is nothing to be gained, neither in efficiency nor in simplicity. So we decided not to create per processor files but only one file where all processors write to by parallel output utilizing MPI IO. A nice thing is that this way we get restart on a different number of processors for free.

Of course on systems that do not support globally shared disc space we have to fall back on the inferior approach of creating many files and merging them later on.

Splitting of Huge Data Files

Large calculations with many time steps produce output file sizes that are inconvenient to say the least. To be able to calculate an unlimited number of time steps we have to split our output files. This can be very easily done because the control file tells which binary files to use, anyway. So we only need to point to a new set of binary files at the right place inside the control file and the filters will know where to find the result data.

---

[4] There is no sorting algorithm involved here. We only need to maintain the ordering with little extra effort.

File Format Summary

Figure 7 shows a schematic sketch of our file format. The plain text control file is symbolized by the hierarchical structure on the left side. For each physical field in the calculation and each result there is a description in the control file. These descriptions contain the offsets where the associated result values are stored in the binary data files. There are two binary files, one contains four byte integers and the other one double precision floating point numbers. The binary files, depicted in the middle, consist of consecutive result blocks called chunks. These chunks are made of entries of constant length, there is one entry per element or node.

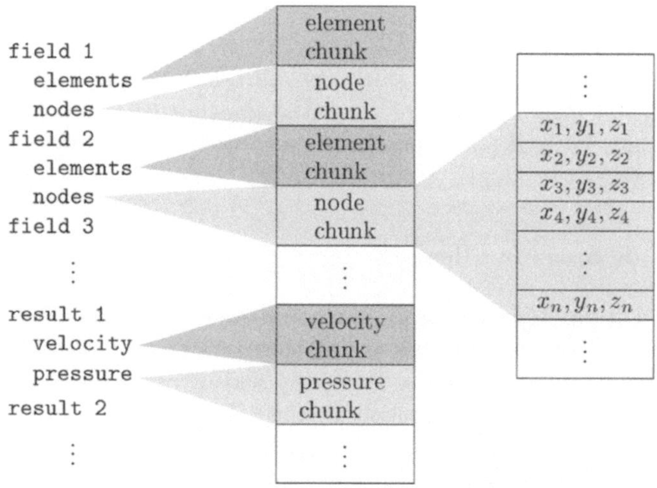

Fig. 7. Sketch of our file format: The plain text control file describes the structure (left), the binary files consist of consecutive chunks (middle) and each chunk contains one entry for each element or node (detail on the right side)

4.4 Algorithmic Details

It is the details that matter. The little detail that needs further consideration is how we are going to produce the described kinds of files.

One Write Operation per Processor for Each Result Chunk

Because the files are written in parallel each processor must write its part of the result with just one MPI IO call. Everything else would require highly sophisticated synchronization and thus create both a communication disaster and an efficiency nightmare. But this means that each processor writes a consecutive piece of the result and in particular, because of our ordering by id, each processor writes the result values from a consecutive number of nodes or elements.

The original distribution of elements and nodes to the processors, however, follows a very different pattern that is guided by physical considerations. It is not guaranteed that elements and nodes with consecutive ids live on the same processor. That means in order to write one result chunk with one MPI IO call we have to communicate each result value from the processor that calculated it to the one that will write this value. In general each processor must communicate with any other to redistribute the result values, but each pair of processors needs to exchange a distinct set of values. J.G. Kennedy et al. [15] describe how to do this efficiently.

The key point is that each processor needs as many send operations as there are processors participating in order to distribute its values. Likewise each processor needs as many receive operations as there are processors involved. Now the `MPI_Sendrecv` function allows to send values to one processor and at the same time receive values from another one. Using that function it is possible to interweave the sends and receives so that all processors take part in each communication step and the required number of communication steps only equals the number of participating processors. Figure 8 shows the communication pattern with four participating processors $P1$ to $P4$. In this case it takes four communication steps, a to d, to redistribute the result values completely. The last communication step, however, is special, there each processor communicates with itself. This is done for convenience only, otherwise we would need a special treatment for those result values that are calculated and written by the same processor.

This redistribution has to be done with every result chunk, it is an integral part of the output algorithm. However, because both distributions, the physical distribution used for calculation as well as the output distribution, are fixed when the calculation starts, it is possible to set up the redistribution information

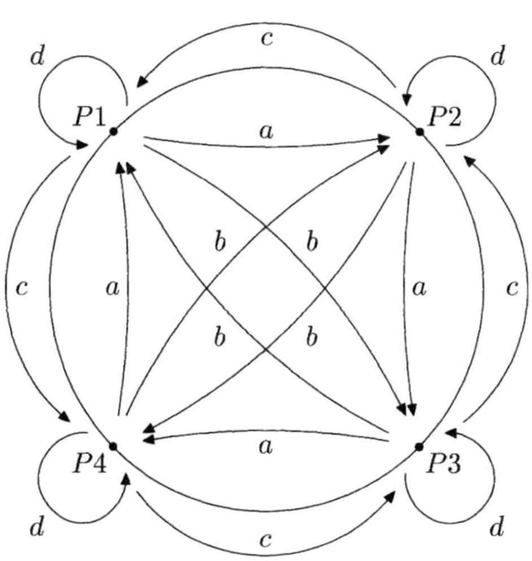

Fig. 8. The communication that redistributes the result values for output between four processors needs four steps labeled a, b, c and d

before the calculation begins. In particular the knowledge which element and node values need to be send to what processor can be figured out in advance.

The redistribution pattern we use is simple but not the most efficient possible. In particular neither hardware depended processor distances nor variable message length are taken into account. Much more elaborated algorithms are available, see Guo and Pan [16] and references presented there. In parallel, highly nonlinear Finite Element calculations, however, output is not performance dominating. That is why a straight parallel implementation is preferred over more complex alternatives.

5 Conclusion

In the present paper several aspects of computational efficiency of parallel finite element simulations were addressed.

In the first part a straightforward approach for a very efficient implementation of the element calculations for advanced finite elements on unstructured grids has been discussed. This concept, requiring only little changes to an existing code, achieved a high performance on the intended vector architecture and also showed a good improvement in the efficiency on other platforms. By grouping computationally similar elements together the length of the innermost loop can be controlled and adapted to the current hardware. In addition the effect of different programming languages and different array management techniques on the performance was investigated.

The main bulk of the numerical work, the solution of huge systems of linear equations, speeds up a lot with the appropriate sparse matrix format. Diagonal sparse matrix storage formats win over row or column formats in the case of vector machines because they lead to long vector lengths. Block computations are necessary to achieve a good portion of peak performance not only on vector machines, but also on most of the superscalar architectures. Block oriented preconditioning techniques are considered numerically superior to the point oriented ones.

The introduced parallel IO subsystem provides a platform independent, flexible yet efficient way to store and retrieve simulation results. The output operation is fully parallel and well scalable to a large number of processors. The number of output operations is kept to a minimum, keeping performance penalties induced by hard disc operation low. Reliability considerations are addressed by unstructured, accessible binary data files along with human readable plain text structure information.

Acknowledgements

The authors would like to thank Uwe Küster of the ‘High Performance Computing Center Stuttgart’ (HLRS) for his continuing interest and most helpful advice and the staff of ‘NEC – High Performance Computing Europe’ for the constant technical support.

References

1. Behr, M., Pressel, D.M., Sturek, W.B.: Comments on CFD Code Performance on Scalable Architectures. Computer Methods in Applied Mechanics and Engineering **190** (2000) 263–277
2. Oliker, L., Canning, A., Carter, J., Shalf, J., Skinner, D., Ethier, S., Biswas, R., Djomehri, J., van der Wijngaart, R.: Evaluation of Cache-based Superscalar and Cacheless Vector Architectures for Scientific Computations. In: Proceedings of the ACM/IEEE Supercomputing Conference 2003, Phoenix, Arizona, USA. (2003)
3. Veldhuizen, T.L.: Scientific Computing: C++ Versus Fortran: C++ has more than caught up. Dr. Dobb's Journal of Software Tools **22** (1997) 34, 36–38, 91
4. Veldhuizen, T.L., Jernigan, M.E.: Will C++ be Faster than Fortran? In: Proceedings of the 1st International Scientific Computing in Object-Oriented Parallel Environments (ISCOPE'97). Lecture Notes in Computer Science, Springer-Verlag (1997)
5. Pohl, T., Deserno, F., Thürey, N., Rüde, U., Lammers, P., Wellein, G., Zeiser, T.: Performance Evaluation of Parallel Large-Scale Lattice Boltzmann Applications on Three Supercomputing Architectures. In: Proceedings of the ACM/IEEE Supercomputing Conference 2004, Pittsburgh, USA. (2004)
6. Ethier, C., Steinman, D.: Exact Fully 3d Navier Stokes Solution for Benchmarking. International Journal for Numerical Methods in Fluids **19** (1994) 369–375
7. Wall, W.A.: Fluid-Struktur-Interaktion mit stabilisierten Finiten Elementen. phdthesis, Institut für Baustatik, Universität Stuttgart (1999)
8. D'Azevedo, E.F., Fahey, M.R., Mills, R.T.: Vectorized Sparse Matrix Multiply for Compressed Row Storage Format. In: Proceedings of the 5th International Conference on Computational Science, Atlanta, USA. (2005)
9. Tuminaro, R.S., Shadid, J.N., Hutchinson, S.A.: Parallel Sparse Matrix Vector Multiply Software for Matrices with Data Locality. Concurrency: Practice and Experience **10-3** (1998) 229–247
10. Nakajima, K.: Parallel Iterative Solvers of GeoFEM with Selective Blocking Preconditioning for Nonlinear Contact Problems on the Earth Simulator. GeoFEM 2003-005, RIST/Tokyo (2003)
11. National Center for Supercomputing Applications. University of Illinois: Hierarchical Data Format. http://hdf.ncsa.uiuc.edu (2005)
12. Unidata Community: Network Common Data Form. http://my.unidata.ucar.edu/content/software/netcdf/index.html (2005)
13. Hunt, A., Thomas, D.: The Pragmatic Programmer: From Journeyman to Master. Addison-Wesley, Reading, MA (2000)
14. Aho, A.V., Sethi, R., Ullman, J.D.: Compilers. Addison-Wesley, Reading, MA (1986)
15. Kennedy, J., Behr, M., Kalro, V., Tezduyar, T.: Implementation of implicit finite element methods for incompressible flows on the CM-5. Computer Methods in Applied Mechanics and Engineering **119** (1994) 95–111
16. Guo, M., Pan, Y.: Improving Communication Scheduling for Array Redistribution. Journal of Parallel and Distributed Computing **(5)65** (2005) 553–563

The Role of Supercomputing in Industrial Combustion Modeling

Natalia Currle-Linde[1], Benedetto Risio[2], Uwe Küster[1], and Michael Resch[1]

[1] High Performance Computing Center Stuttgart (HLRS),
Nobelstraße 19, D-70569 Stuttgart, Germany,
linde@hlrs.de,
[2] RECOM Services, Nobelstraße 15, D-70569 Stuttgart, Germany

Abstract Currently, numerical simulation using automated parameter studies is already a key tool in discovering functional optima in complex systems such as biochemical drug design and car crash analysis. In the future, such studies of complex systems will be extremely important for the purpose of steering simulations. One such example is the optimum design and steering of computation equipment for power plants. The performance of today's high performance computers enables simulation studies with results that are comparable to those obtained from physical experimentation. Recently, Grid technology has supported this development by providing uniform and secure access to computing resources over wide area networks (WANs), making it possible for industries to investigate large numbers of parameter sets using sophisticated optimization simulations. However, the large scale of such studies requires organized support for the submission, monitoring, and termination of jobs, as well as mechanisms for the collection of results, and the dynamic generation of new parameter sets in order to intelligently approach an optimum. In this paper, we describe a solution to these problems which we call Science Experimental Grid Laboratory (SEGL). The system defines complex workflows which can be executed in the Grid environment, and supports the dynamic generation of parameter sets.

1 Introduction

During the last 20 years the numerical simulation of engineering problems has become a fundamental tool for research and development. In the past, numerical simulations were limited to a few specified parameter settings. Expensive computing time did not allow for more. More recently, high performance computer clusters with hundreds of processors enable the simulation of complete ranges of multi-dimensional parameter spaces in order to predict an operational optimum for a given system. Testing the same program in hundreds of individual cases may appear to be a straightforward task. However, the administration of a large number of jobs, parameters and results poses a significant problem. An effective mechanism for the solution of such parameter problems can be created using the resources of a Grid environment. This paper, furthermore proposes the coupling of these Grid resources to a tool which can carry out the following: generate parameter sets, issue jobs in the Grid environment, control the successful operation and termination of these jobs, collect results, inform the user about ongoing work and generate new parameter sets based on previous results in order to approach a functional optimum, after which the mechanism should gracefully terminate. We expect to see the use of parameterized simulations in many disciplines. Examples are drug design, statistical crash simulation of cars, airfoil design, power plant simulation. The mechanism proposed here offers a unified framework for such large-scale optimization problems in design and engineering.

1.1 Existing Tools for Parameter Investigation Studies

Tools like Nimrod [1] and Ilab [1] enable parameter sweeps and jobs, running them in a distributed computer environment (Grid) and collecting the data. ILab also allows the calculation of multi-parametric models in independent separate tasks in a complicated workflow for multiple stages. However, none of these tools is able to dynamically generate new parameter sets by an automated optimization strategy. In addition to the above mentioned environments, tools like Condor [1], UNICORE [2] or AppLeS [1] can be used to launch pre-existing parameter studies using distributed resources. These, however, give no special support for dynamic parameter studies.

1.2 Workflow

Realistic application scenarios become increasingly complex due to the necessary support for multiphysics applications, preprocessing steps, postprocessing filters, visualization, and the iterative search in the parameter space for optimum solutions. These scenarios require the use of various computer systems in the Grid, resulting in complex procedures best described by a workflow specification. The definition and execution of these procedures requires user-friendly workflow description tools with graphical interfaces, which support the specification of loops, test and decision criteria, synchronization points and communication via messages. Several Grid workflow systems exist. Systems such as Triana [3] and UNICORE, which are based on directed acyclic graphs (DAG), are limited with respect to the power of the model; it is difficult to express loop patterns, and the expression of process state information is not supported. On the other hand, workflow-based systems such as GSFL [4], and BPEL4WS [4] have solved these problems but are too complicated to be mastered by the average user. With these tools, even for experienced users, it is difficult to describe nontrivial workflow processes involving data and computing resources. The SEGL system described here aims to overcome these deficiencies and to combine the strengths of Grid environments with those of workflow oriented tools. It thus provides a visual editor and a runtime workflow engine for dynamic parameter studies.

1.3 Dynamic Parameterization

Complex parameter studies can be facilitated by allowing the system to dynamically select parameter sets on the basis of previous intermediate results. This dynamic parameterization capability requires an iterative, self-steering approach. Possible strategies for the dynamic selection of parameter sets include genetic algorithms, gradient-based searches in the parameter space, and linear and nonlinear optimization techniques. An effective tool requires support of the creation of applications of any degree of complexity, including unlimited levels of parameterization, iterative processing, data archiving, logical branching, and the synchronization of parallel branches and processes. The parameterization of data is an extremely difficult and time-consuming process. Moreover, users are very sensitive to the level of automation during application preparation. They must be able to define a fine-grained logical execution process, to identify the position in the input data of parameters to be changed during the course of the experiment, as well as to formulate parameterization rules. Other details of the parameter study generation are best hidden from the user.

1.4 Databases

The storage and administration of parameter sets and data for an extensive parameter study is a challenging problem, best handled using a flexible database. An adequate database capability must support the *a posteriori* search for specific behavior not anticipated in the project. In SEGL the automatic creation of the project and the administration of data are based on an object-oriented database (OODB) controlled by the user. The database collects all relevant information for the realization of the experiment, such as input data for the parameter study, parameterization rules and intermediate results. In this paper we present a concept for the design and implementation of SEGL, an automated parametric modeling system for producing complex dynamically-controlled parameter studies.

2 System Architecture and Implementation

Figure 1 shows the system architecture of SEGL. It consists of three main components: the User Workstation (Client), the ExpApplicationServer (Server) and the ExpDBServer (OODB). The system operates according to a Client-Server-Model in which the ExpApplication Server interacts with remote target computers using a Grid Middleware Service. The implementation is based on the Java 2 Platform Enterprise Edition (J2EE) specification and JBOSS Application Server. The System runs on Windows as well as on UNIX platforms. The OODB is realized using the Java Data Objects (JDO) implementation of Fast Objects [5].

The client on the user's workstation is composed of the ExpDesigner and the ExpMonitorVIS. The ExpDesigner is used to design, verify and generate the experiment's program, organize the data repository and prepare the initial data. The ExpMonitorVIS is generated for visualization and for the actual control of the complete process. The ExpDesigner allows to describe complex experiments using a simple graphical language. Each experiment is described at three levels: control flow, data flow and data repository. The control flow level is used for the description of the logical schema of the experiment. On this level the user makes a logical connection between blocks: direction, condition and sequence of the execution of blocks. Each block can be represented as a simple parameter study.

The data flow level is used for the local description of interblock computation processes. The description of processes for each block is displayed in a new window. The user is able to describe:

(a) Both a standard computation module and a user-specific computation module. The user-specific module can be added to suit the application domain.
(b) The direction of input and output data between the metadata repository and the computation module.
(c) The parameterization rules for the input set of the data.
(d) Synchronization of interblock processes.

On the data repository level, a common description of the metadata repository is created. The repository is an aggregation of data from the blocks at the data flow level. Each block contains one or more windows representing part of the data flow. Also described at the data repository level are the key and service fields (objects) of the database.

After completion of the design of the program at the graphical icon-level, it is "compiled". During the "compilation" the following is created:

(a) a table of the connections between program objects on the data flow level for each block (manipulation of data) and
(b) a table of the connections between program blocks on the control flow level for the experiment.

Parallel to this, the experiment's database aggregates the data base icon objects from all blocks/windows at the data flow level and generates query-language (QL) descriptions of the experiment's database. The container application of the experiment is transferred to the ExpApplicationServer and the QL descriptions are transferred to the server database. Here, the metadata repository is created. The ExpApplicationServer consists of the ExpEngine, the Task, the ExpMonitorSupervisor and the ResourceMonitor.

The Task is the container application. The ResourceMonitor holds information about the available resources in the Grid environment. The MonitorSupervisor controls the work of the runtime system and informs the Client about the current status of the jobs and the individual processes. The ExpEngine is the controlling subsystem of SEGL (Runtime subsystem). It consists of three subsystems: the TaskManager, the JobManager and the DataManager. The TaskManager is the central dispatcher of the ExpEngine coordinating the work of the DataManager and the JobManager:

(1) It organizes and controls the sequence of execution of the program blocks. It starts the execution of the program blocks according to the task flow and the condition of the experiment program.

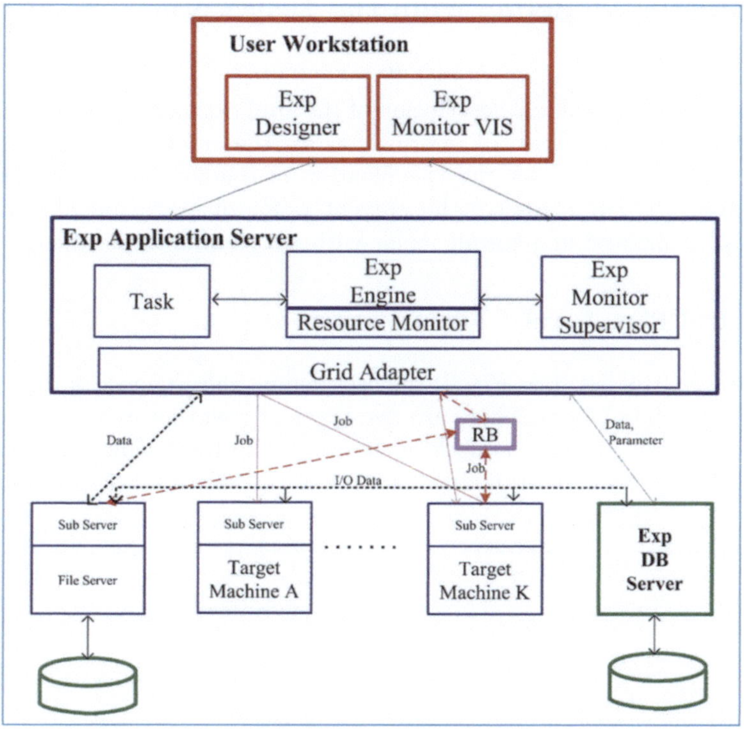

Fig. 1. System Architecture

(2) It activates a particular block according to the task flow, chooses the necessary computer resources for the execution of the program and deactivates the block when this section of the program has been executed.
(3) It informs the MonitorSupervisor about the current status of the program.

The DataManager organizes data exchange between the ExpApplicationServer and the FileServer and between the FileServer and the ExpDBServer. Furthermore, it controls all parameterization processes of input data. The JobManager generates jobs and places them in the corresponding SubServer of the target machines. It controls the placing of jobs in the queue and observes their execution.

The final component of SEGL is the database server (ExpDBServer). All data which occurred during the experiment, initial and generated, are kept in the ExpDBServer. The ExpDBServer also hosts a library tailored to the application domain of the experiment. For the realization of the database we choose an object-oriented database because its functional capabilities meet the requirements of an information repository for scientific experiments. The interaction between ExpApplicationServer and the Grid resources is done through a Grid Adaptor. Currently, e.g. Globus [6] and UNICORE offer these services.

3 Parameter Modeling from the User's View

Figure 2 shows an example of a task flow for an experiment as it appears in the ExpDesigner. The graphical description of the application flow has two purposes: firstly, it is used to collect all information for the creation of the experiment and, secondly, it is used for the visualization of the current experiment in the ExpMonitorVIS. For instance, the current point of execution of a computer process is highlighted in a specific color within a running experiment.

3.1 Control Flow Level

Within the control flow (see Fig. 2) the user defines the sequence of the execution of the experiments blocks. There are two types of operation block: control block and solver block. The solver block is the program object which performs some complete operation. The standard example of the solver block can be a simple

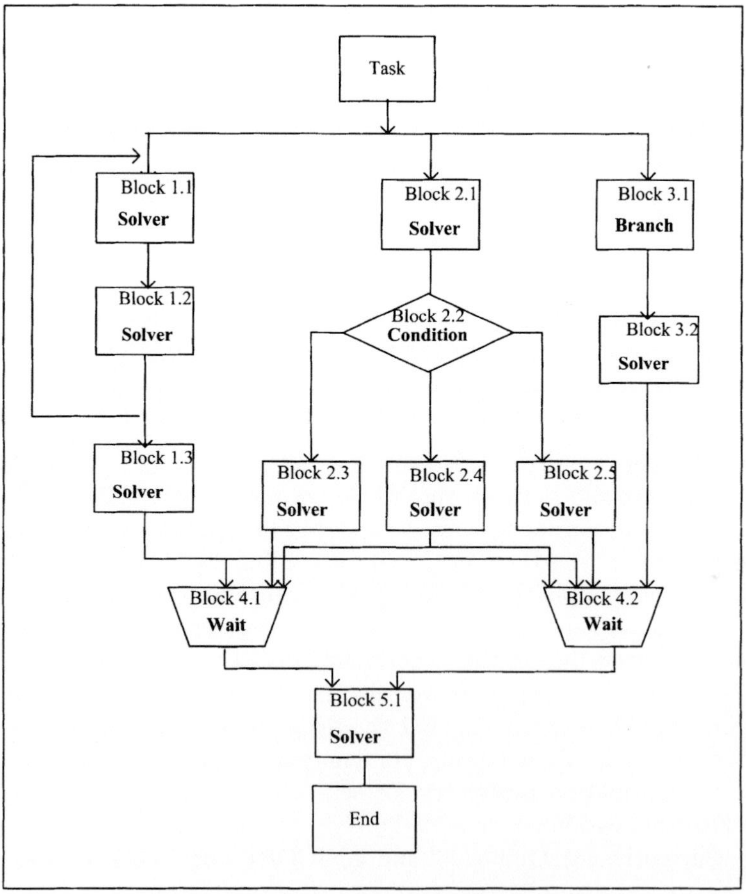

Fig. 2. Sample Task Flow (control flow)

parameter sweep. The control block is the program object which allows the changing of the sequence of execution operation according to a specified criterion.

Figure 2 shows an example of task flow. After execution of "Task" block 1.1, block 2.1 and block 3.1 are activated simultaneously. In each of these blocks a process is executed. After having worked with the first set of data in block 1.1, the first process in block 1.2 is activated. After execution of the first process in block 1.2, the first process in block 1.3 and the second process in block 1.1 are started according to the logic of the experiment. The input data for the second and the following processes in block 1.1 are prepared in block 1.2 and so on.

3.2 Data Flow Level

Figure 3 presents an example of a solver block (Block 1.1). At this level, the user can describe the manipulation of data in a very fine grained way. The solver block consists of computation (C), replacement (R), parameterization (P) modules and a database. These are connected to each other with arrowed lines showing the direction of data transfer between modules and the sequence of execution during the computation process.

Each module is a Java object, which has a standard structure and consists of several sections. For example: each computation module (C) consists of four sections. The first section organizes the preparation of input data. The second generates the job and controls its execution. The third initializes and controls the record of the result in the experiment database. The fourth section controls the execution of module operation. It also informs the main program of the block about the manipulation of certain sets of data and when execution within a block is complete.

After a block is started, the parameterization module (P) and replacement module (R) wait for the request from the corresponding inputs of the computation module (C). After that, they generate a set of input data according to rules specified by the user, either as mathematical formulae or a list of parameter values. In this example three variants of parameterization are represented:

(a) Direct transmission of the parameter values with the job. In this case, parameterization module (P3) transfers the generated parameter value to the computation module (C1) upon its request. The computation module generates the job, including converting parameter values into corresponding job parameters. This method can be used if the parameterized value is a number, symbol or combination of both.
(b) Parameterized objects are large arrays of information (DB-P4 in Fig. 3) which are kept in the experiment database. These parameters are copied directly from the experiment database to the corresponding file server and then written with the same array name with the index of the number of the stage. In this case, attributes of the job are sent to the file server as references (an array of data).
(c) If it is important, then the preparation of the data is moved outside of the main program. This allows the creation of a more universal computation

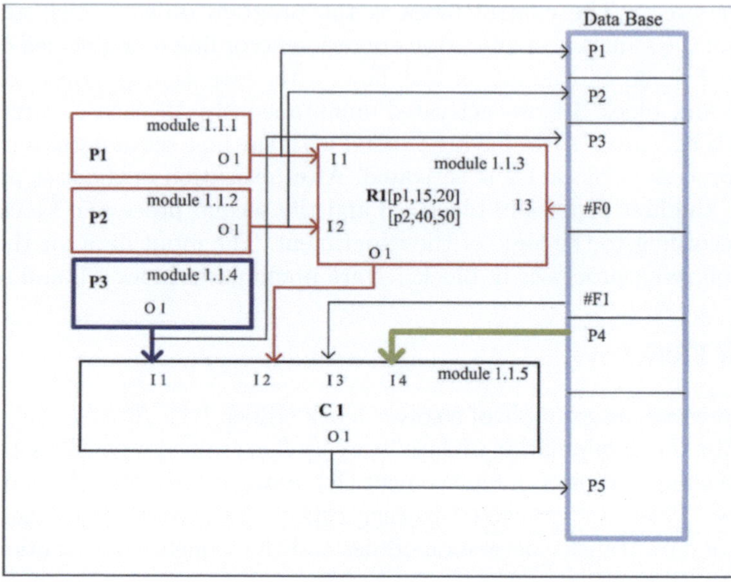

Fig. 3. Solver Block 1.1 (data flow)

module. Furthermore, it allows scaling, i.e. avoiding limitations in the size, position, type and number of the parameterized objects used in a module.

In these cases the replacement module is used. During the preparation of the next set of input data, new parameter values P1 and P2 are generated. The generated parameter set is linked with replacement processes and then delivered to the corresponding FileServer, where the replacement process is executed.

After the replacement of the specified parameters, the input data is ready for the first stage of computation. Computation module C1 sends a message to the JobManager to prepare the job for the first stage. The JobManager chooses the computer resources currently available in the network and starts the job. After confirmation from the corresponding SubServer of the Target Machine that the job is in a queue, the preparation of the next set of data for the next computation stage begins. Each new stage carries out the same processes as the previous stage. At all stages, the output file is archived immediately after being received by the experiment's database. The control of all processes takes place according to the pattern described above. After starting the ExpMonitorVIS on their workstation, the user receives continuously updated status information regarding the experiment's progress.

4 Use case: Power Plant Simulation by Varying Burners and Fuel Quality

The liberalization of the energy markets puts more and more pressure on the competitiveness of power companies throughout the world. In order to maintain their competitive edge, it is necessary to optimize the operation of existing power plants towards minimum operational costs. Potential optimization targets can be minimization of excess air (increasing efficiency) or NOx-emission (reducing DeNOx operation costs). Pure experimental optimizations without computer-aided techniques are time-consuming and require a significantly higher manpower effort. Furthermore, in the case of necessary design changes the technical risks involved in the investment decision can only be assessed with computer-aided techniques. Computer-aided methods are well accepted in the power industry. The optimization procedure applied by SEGL for the present problem is based on a genetic algorithm (GA).

In order to work on boiler optimization problems with SEGL, the parameters that have to be optimized are coded in binary form and assembled to a so-called "chromosome". The chromosome carries all the important properties to be changed of the so-called "individuals". A certain number of these artificial individuals are generated initially, the so-called "population", and the GA of SEGL imitates the natural evolution process. The imitation is done by applying the genetic mechanisms Selection, Recombination and Mutation. An illustration of the basic workflow in the SEGL is shown in Fig. 4.

The basic workflow can be described as follows:

1. Binary coding of optimization parameters and chromosome assembly.
2. Generation of an initial population.
3. Decoding of the chromosome information for each individual.

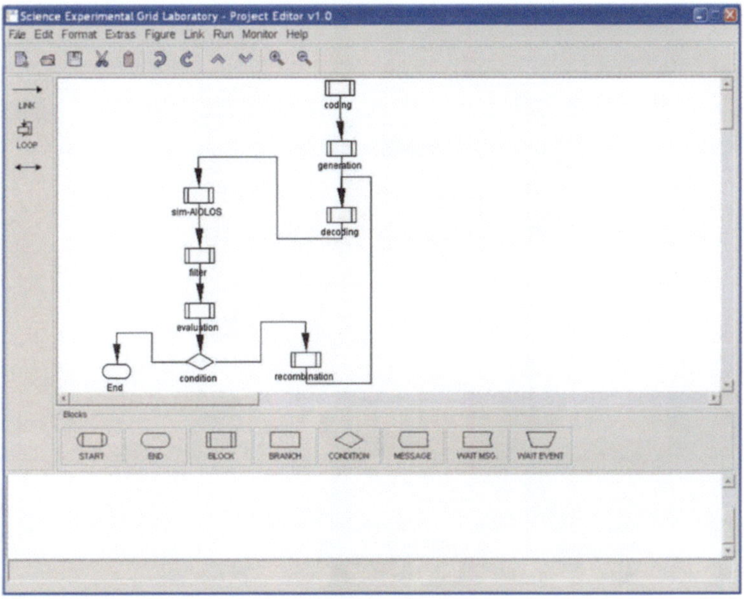

Fig. 4. Workflow

4. Simulation of the decoded set of optimization parameters with the 3D-furnace simulation code RECOM-AIOLOS for each individual. This is the time consuming step.
5. Filtering the 3-D results of the furnace simulation to derive the target values for each individual.
6. Evaluation of the performance level for each individual (terminate the optimization process if desired optimization level is reached).
7. Selection of suitable individuals for reproduction and Recombination/Mutation of the chromosome information for the selected individuals to generate new individuals.
8. Return to Step 3 for new individuals.

4.1 Industrial Applicability

An experimental operation optimization exercise performed in 1991 at a power station in Italy (ENEL's coal-fired Fusina) is used to demonstrate the capabilities of SEGL. In a windbox, the amount of air flowing through a nozzle is controlled by the damper setting of the nozzle. A damper setting of 100% means that the flow passage of the nozzle is fully open. Reducing the damper setting of a single nozzle allows for reduction of the air mass flow through the nozzle, but at the same time the air mass flows for all other nozzles in the windbox are increased.

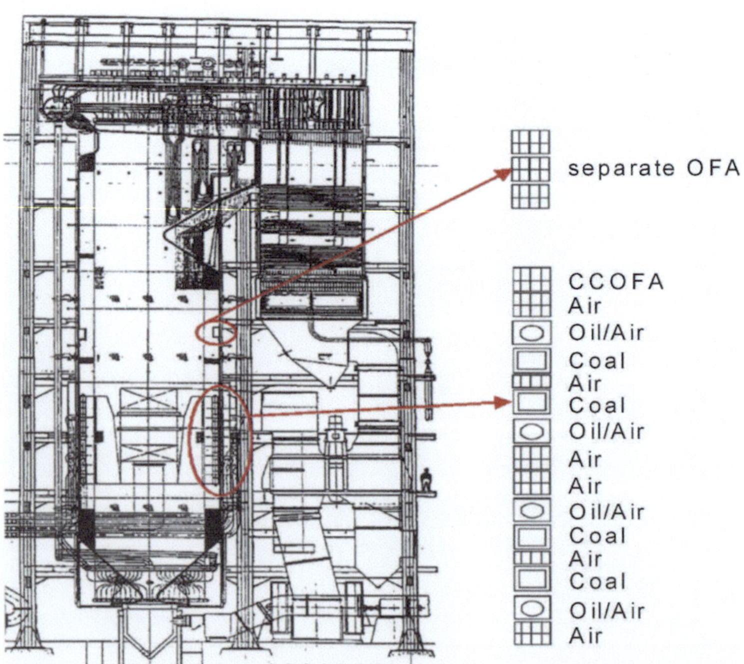

Fig. 5. Firing and separate OFA arrangement fur Fusina #2

In 1991 separate overfire air nozzles (separate OFA) were installed above the main combustion zone (see Fig. 4) to minimize NOx-emissions. A new operation mode was required after the successful installation of the separate overfire air to maintain the lowest possible NOx-emission together with a minimum unburned carbon loss. In 1991 this optimization exercise was solved experimentally. In a series of 15 tests over a duration of approximately 10 days, 15 operation modes were tested with varying amounts of close coupled overfire air (CCOFA), separate OFA, and tilting angle of the separate OFA (±30°).

The following operation experience was recorded to identify an optimized operation:

(a) For a horizontal orientation of the separate OFA the maximum NOx-reduction is reached with dampers 100% open.
(b) A tilting of the separate OFA to −30° has a minor effect on the NOx-emission but improves the burnout (reduced unburned carbon loss).
(c) A tilting of the separate OFA to +30° leads to an NOx-reduction but increases the unburned carbon loss significantly.
(d) Closing the CCOFA completely at 100% open separate OFA has only a minor effect on the NOx-emission.

In order to work on this combustion optimization problem in virtual reality, a high-resolution boiler model with 1 million grid points was generated. As shown in Table 1, an accuracy of approximately ±10% between simulation and reality can be reached on the high-resolution boiler model. The optimization parameters "OFA damper setting", "CCOFA damper setting", and "Tilting Angle"

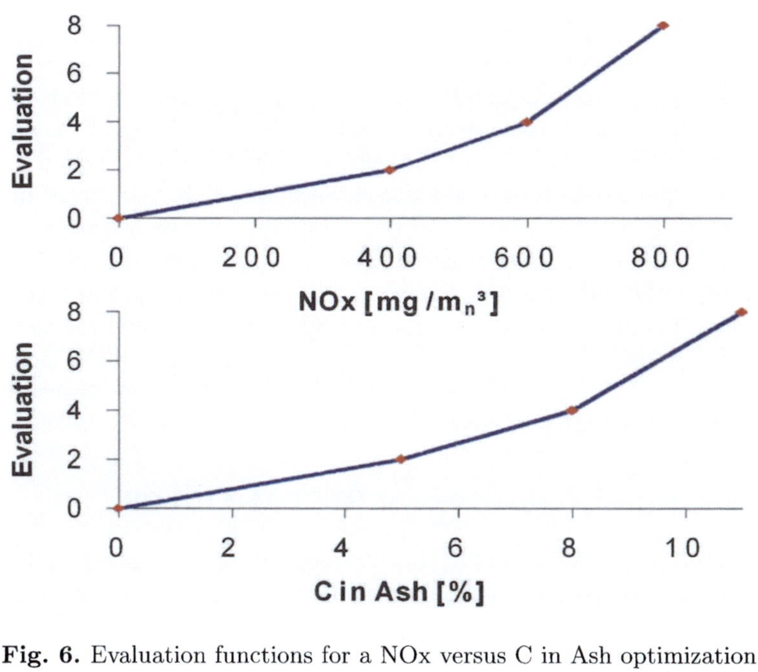

Fig. 6. Evaluation functions for a NOx versus C in Ash optimization

Table 1. Measured and calculated NOx-emission and C in Ash

	NOx-emission [mg/m$_n^3$, 6% O_2]		C in Ash [%]	
Setting	measured	calculated	measured	calculated
No OFA No CCOFA	950–966	954	6.41–7.50	5.66
No OFA CCOFA: 100%	847–858	794	7.47–7.61	6.58
OFA:100% CCOFA: 100%	410–413	457	10.43–11.48	10.28

Table 2. Development of best individuals in each generation during automatic optimization

Generation	Target-Value	OFA [%]	CCOFA [%]	Tilting Angle [°]	NOx mg/m$_n^3$	C in Ash [%]
Basis	12.070	0	0	0	805	3.39
1	10.061	100	100	−30	479	10.84
5	9.600	93	93	−30	473	10.42
10	9.177	93	20	−30	458	10.26

were coded with 4 bit on the chromosomes. NOx-emission and C in Ash values achieved in the model were combined to a target function for the evaluation of the individuals. The underlying combined evaluation target function are shown in Fig. 6.

$$Target\ Function = Evaluation[\text{NOx}] + Evaluation[C\ \ in\ \ Ash]$$

The GA required approximately 11 generations with 10 individuals per population to identify an optimized parameter set. During the course of the automatic optimization, approximately 51 of the entire 4096 ($2^4 \cdot 2^4 \cdot 2^4$) coded combinations of parameter settings were evaluated with respect to the target functions. Table 2 shows the development of the best individuals in each generation in the course of the automatic optimization. The results demonstrate that SEGL is able to identify the same positive measures that were found in the experimental optimization. The final run on the high-resolution boiler model led to an NOx-emission of 476 mg/m$_n^3$ at 6% O_2 and a C in Ash value of 8.42%. Both values are in the range of the emission and C in Ash values that were observed in the field after the optimization exercise.

4.2 Computational Performance of RECOM-AIOLOS

As well as accuracy, investigated in the previous section, computational economy is an important requirement in the industrial use of 3D-combustion simulations. The aim is to obtain solutions of acceptable accuracy within short time periods and at low financial costs.

Table 3. Computational performance on varying number of processors and problem size

Problem size	Processors	Gas combustion	Solid Fuel combustion
5 Mio. Grid points	1 processor	6.3 GFlops	4.3 GFlops
1 Mio. Grid points	1 node=8 processors	24.9 GFlops	17.2 GFlops
5 Mio. Grid points	1 node=8 processors	30.7 GFlops	21.2 GFlops
10 Mio. Grid points	1 node=8 processors	36.4 GFlops	25.1 GFlops
10 Mio. Grid points	4 node=64 processors	122.2 GFlops	84.3 GFlops

In order to exploit the possibilities of parallel execution RECOM-AIOLOS has successfully been parallelized in the past with two different strategies: a domain decomposition method using MPI (Message Passing Interface) as the message passing environment [7] and a data parallel approach using Microtasking [8]. These investigations were performed either on distributed memory massively parallel computers (MPPs) or pure shared memory vector computers (PVPs), showing acceptable parallel efficiencies for both approaches.

The architecture used in the present paper is a 72-node NEC SX-8 with an aggregate peak-performance of 12 TFlops and a shared main memory of 9.2 TB. The NEC SX-8 supports a hybrid parallel programming model that allows combination of distributed memory parallelization across nodes and data parallel execution with the node.

The degree of vectorization of AIOLOS hereby defined as the ratio between the time spent in the vector unit and the total user time is greater than 99.7% depending on the problem size.

Table 3 shows the computational performance on varying number of processors and problem size. The results indicate that the code achieves 39% of the theoretical single processor peak performance of 16 GFlops for the gas combustion model. In the case of the solid fuel combustion model, only 27% of the single processor peak performance is reached.

The total duration of the automatic optimization described in the previous chapter was 3 days. The total optimisation consumed 581 CPUh.

5 Conclusion

This paper presented the concept and description of the implementation of SEGL for the design of complex and hierarchical parameter studies which offers an efficient way to execute scientific experiments. We can show that SEGL allows for substantial reduction in optimization costs for parameter studies.

This is a prerequisite for applying automatic optimization techniques to industrial combustion problems that will require hundreds of variations to be run within today's project time frames to derive practical conclusions for industrial combustion equipment. High performance computers are helpful for this purpose but high aggregated machine performance alone is not enough. Tools will be needed for managing virtual tests and the immense amount of data the simulations produce. This will allow for an automated data handling and post-processing.

References

1. de Vivo, A., Yarrow, M., McCann, K.: A comparison of parameter study creation and job submission tools. Technical report, NASA Ames Research Center (2000)
2. Erwin, D.E.: Joint project report for the BMBF project UNICORE plus. Grant Number: 01 IR 001 A-D, Duration: January 2000 – December 2002 (2003)
3. Taylor, I., Shields, M., Wangand, I., Philp, R.: Distributed P2P computing within triana: A galaxy visualization test case. In: IPDPS 2003 Conference. (2003)
4. Tony, A., Curbera, F., Dholakia, H., Goland, Y., Klein, J., Leymann, F., Liu, K., Roller, D., Smith, D., Thatte, S., Trickovic, I., Weerawarana, S.: Specification: Business process execution language for web services version 1.1. Technical report, NASA Ames Research Center (2003)
5. Corporation, V.: Fastobject webpage. http://www.fastobjects.com (2005)
6. Foster, I., Kesselman, C.: The globus project: A status report. In: Proc. IPPS/SPDP '98 Heterogeneous Computing Workshop. (1998)
7. Lepper, J., Schnell, U., Hein, K.R.G.: Numerical simulation of large-scale combustion processes on distributed memory parallel computers using mpi. In: Parallel CFD 96. (1996)
8. Risio, B., Schnell, U., Hein, K.R.G.: HPF-implementation of a 3D-combustion code on parallel computer architectures using fine grain parallelism. In: Parallel CFD 96. (1996)

Applications II: Fluid Dynamics

Simulation of the Unsteady Flow Field Around a Complete Helicopter with a Structured RANS Solver

Thorsten Schwarz, Walid Khier, and Jochen Raddatz

German Aerospace Center (DLR),
Member of the Helmholtz Association,
Institute of Aerodynamics and Flow Technology,
Lilienthalplatz 7, D-38108 Braunschweig, Germany
`thorsten.schwarz@dlr.de`
WWW home page: `http://www.dlr.de/as`

Abstract The air flow past a wind tunnel model of an Eurocopter BO-105 fuselage, main rotor and tail rotor configuration is simulated by solving the time dependent Navier-Stokes equations. The flow solver uses overlapping, block structured grids to discretize the computational domain. The simulation setup and the execution on a parallel NEC SX-6 vector computer are described. The numerical results are compared with unsteady pressure measurements on the fuselage and the blades. An overall good agreement is found. Differences between predicted and measured data on the main rotor and the tail rotor can be explained by blade elasticity effects and a different trim law respectively. The computational performance of the flow solver is analyzed for the NEC SX-6 and NEC SX-8 vector computer showing a good parallel performance. Modifications of the code structure resulted in a reduction of the execution time for the Chimera procedure by a factor of 6.6.

1 Introduction

The numerical simulation of the flow around a complete helicopter by solving the unsteady Reynolds-averaged Navier-Stokes (RANS) equations is a challenge. This is mainly due to a lack of available computer resources. The complex flow topology around the helicopter and the unsteadiness of the flow requires computations on grids with millions of grid cells and several thousand physical time steps to solve the governing equations. Only today's supercomputers are fast enough and have enough memory to enable these kind of simulations within a research context. Another issue for helicopter simulations is fluid modeling, e.g. vortex capturing and turbulence modeling.

The flow field around a helicopter is depicted in Fig. 1. A helicopter usually operates at flight speeds below $M = 0.3$. Therefore, the flow is incompressible except for the regions near the blade tips of the main and tail rotor where the

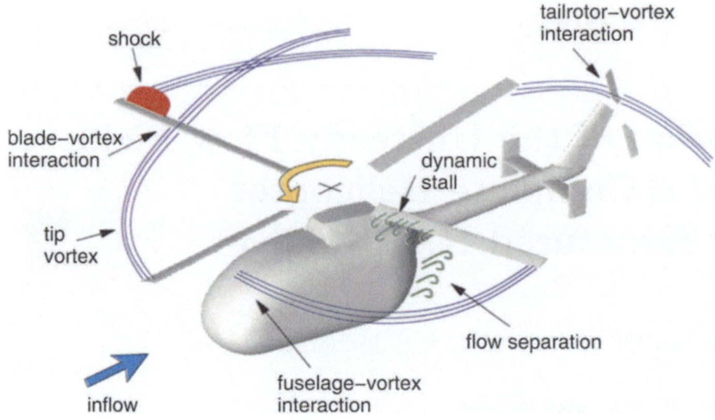

Fig. 1. Aerodynamics of the helicopter

flow may be locally supersonic and shocks may be present. Strong vortices are shed from the blade tips and move downstream with the inflow velocity. These vortices can interact with the following blades. The viscosity of the fluid leads to boundary layers on surfaces and wake sheets downstream of the surfaces. The boundary layers may separate at bluff body components. Flow separation may also occur at the retreating rotor blades, where due to trim considerations the blade incidence angle must be high. Additionally, interactions take place between the helicopter's components, e.g. between the main-rotor, the tail-rotor and the fuselage. All the aforementioned phenomena affect the flight performance of the helicopter, its vibration and its noise emission.

Since flow simulations for complete helicopters are not possible in an industrial environment, the solution of the Navier-Stokes equations is often restricted to individual components of a helicopter. Examples are steady flow simulations for isolated fuselages [1] or unsteady simulations for isolated main rotors [2, 3, 4]. Interactional phenomena between the rotors and the fuselage have been investigated with steady flow simulations, where the main and tail rotors are replaced by actuator discs [5]. The latter are used to prescribe the time averaged effects of the rotors. First Navier-Stokes computations for a full helicopter configuration have been presented in [6, 7, 8].

In an effort to provide the French-German helicopter manufacturer Eurocopter with simulation tools capable of computing the viscous flow around complete helicopters, the project CHANCE [9, 10] was initiated in 1999. Project partners have been the German and French research centers DLR and ONERA, the university of Stuttgart and the helicopter manufacturer Eurocopter. Within the CHANCE project, the flow solvers of DLR and ONERA have been widely extended and were validated for helicopter flows. One final milestone of the project was to simulate the unsteady flow for a complete helicopter configuration. The aim of this paper is to present results obtained by DLR with the block-structured flow solver FLOWer for such a configuration.

2 Simulated Test Case and Flow Conditions

The computations reported here simulate a forward flight test case of a 1:2.5 scale wind tunnel model of an Eurocopter BO-105. The wind tunnel experiment was performed within the EU project HeliNOVI [11] in 2003. (Please note, that most of the HeliNOVI experiments were performed during a second campaign in 2004). Figure 2 shows the model mounted on a model support inside the German-Dutch wind tunnel (DNW). The BO-105 wind tunnel model has a main rotor diameter of 4 m and a tail rotor diameter equal to 0.773 m. Both the main and tail rotors have square blades. The main rotor blades consist of $-8°$ linearly twisted NACA 23012 profile with a chord length equal to 0.121 m. The tail rotor is made of a MBB S 102 E airfoil with zero twist and has a chord length equal to 0.0733 m. All intake and ventilation openings were closed in the experimental model. A cylindrical strut was used to support the model in the wind tunnel. The experimental model, its instrumentation and the wind tunnel tests are described in detail by [12].

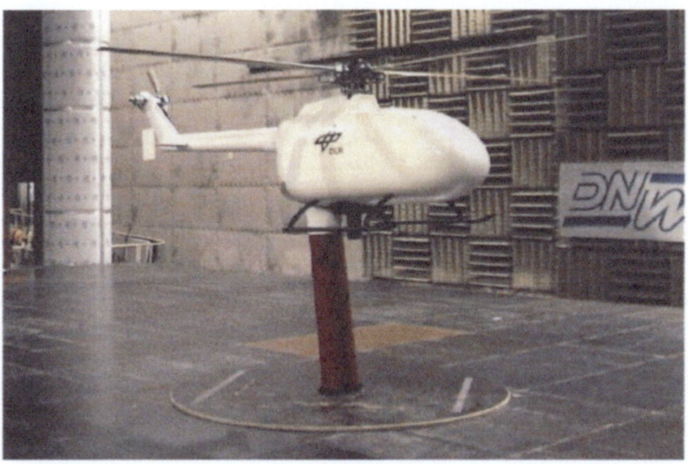

Fig. 2. BO-105 wind tunnel model

The selected test case refers to a forward flight condition with 60 m/s ($M = 0.177$) at an angle of attack equal to $5.2°$. The main and tail rotor angular velocities are equal to 1085 and 5304 RPM respectively, corresponding to a main rotor tip Mach number $M_{\omega R_{MR}} = 0.652$ and a tail rotor tip Mach number $M_{\omega R_{TR}} = 0.63$. The nominal trim law for the main and tail rotor blade pitch angles used in the experiment was $\Theta_{MR} = 10.5° - 6.3° \sin(\Psi_{MR}) + 1.9° \cos(\Psi_{MR})$ for the main rotor and $\Theta_{TR} = 8.0°$ for the tail rotor. Ψ_{MR} is the azimuth angle of the main rotor. Information on the flapping and elastic blade deformation of the main rotor were not available at the time of the simulation. The same holds for the coupled cyclic pitching/flapping motion of the tail rotor.

3 Numerical Approach

DLR's flow solver FLOWer solves the Reynolds-averaged Navier-Stokes equations with a second order accurate finite volume discretization on structured, multi-block grids. The solution process follows the idea of Jameson [13], who represents the mass, momentum and energy fluxes by second order central differences. Third order numerical dissipation is added to the convective fluxes to ensure numerical stability.

FLOWer contains a large array of statistical turbulence models, ranging from algebraic and one-equation eddy viscosity models to seven-equation Reynolds stress models. In this paper a slightly modified version of Wilcox's two-equation k-ω model is used [14, 15]. Unlike the main flow equations, Roe's scheme is employed to compute the turbulent convective fluxes.

For steady flows, the discretized equations are advanced in time using an explicit five-stage Runge-Kutta method. The solution process makes use of acceleration techniques like local time stepping, multigrid and implicit residual smoothing. Turbulence transport equations are integrated implicitly with a DDADI (diagonal dominant alternating direction implicit) method. For unsteady simulations, the implicit dual time stepping method [16, 17] is applied. FLOWer is parallelized based on MPI and is optimized for vector computers.

A method extensively used within the present work is the Chimera overlapping grid technique [18]. This method allows to discretize the computational domain with a set of overlapping grids, see Fig. 3. In order to establish communication between the grids, data from overlapping grids are interpolated for the cells at the outer grid boundaries. If some grid points are positioned inside solid bodies, these points are flagged and are not considered during the flow simulations. The flagged points form a so called hole in the grid. At the hole fringe, data are interpolated from overlapping grids. A detailed description of the Chimera method implemented in FLOWer is given in [19].

The Chimera technique is used in the present computations because of the following reasons. Firstly, compared to alternative approaches (re-meshing for example), relative motion between the different components of the helicopter

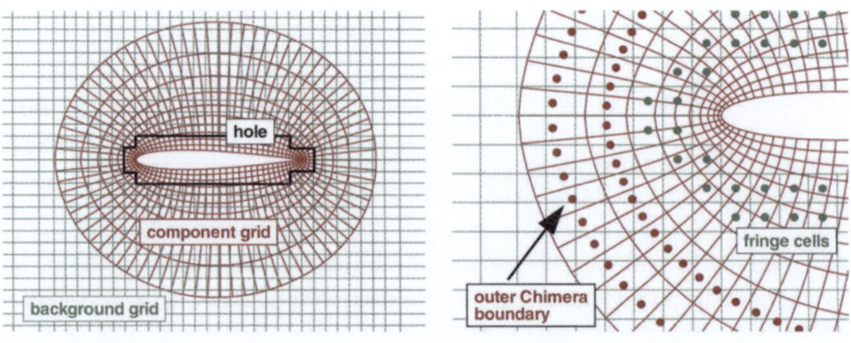

Fig. 3. The Chimera technique, left: overlapping grids, right: interpolation points

can be easily realized. Secondly, Chimera reduces the time and effort required to generate block structured grids around complex configurations.

4 Computational Grid

For the creation of the computational grid, the BO-105 wind tunnel model was subdivided into twelve components: fuselage, left and right stabilizers, four main rotor blades, two tail rotor blades, left and right skids and spoiler with model strut. Multi-block structured grids were generated around each component, see Fig. 4, left. Rotor hubs and drive shafts were not considered in order to simplify mesh generation. Since no wall functions are used, the grids have a high resolution inside the boundary layer. The near field grids were embedded into a locally refined Cartesian background grid with partly anisotropic (non-cubic) cells. A cut through the computational mesh is shown in Fig. 4, right. The interfaces of grid blocks with different cell sizes are realized by patched grids with hanging grid nodes. The automatic grid generator used to create the Cartesian background grid is described in [19]. The complete grid consists of 480 grid blocks

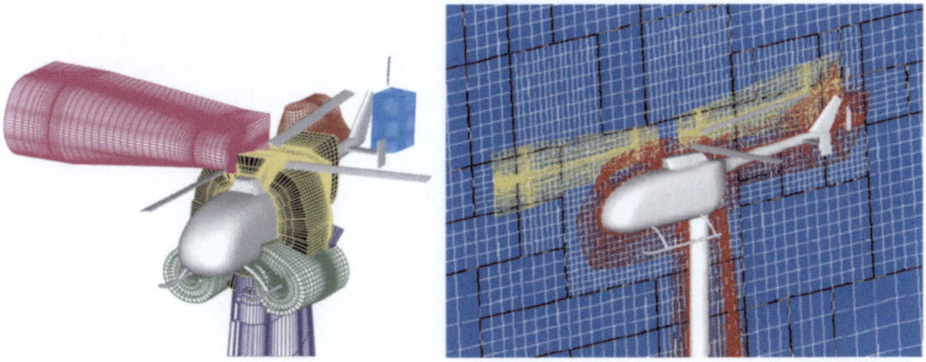

Fig. 4. Computational grid for BO-105 configuration, left: near field grids, right: background grid

Table 1. Grid size

	No. of cells	No. of blocks
fuselage	2171904	17
stabilizer (×2)	734464	5
strut and spoiler	1026048	6
skids (×2)	582912	6
main rotor blade (×4)	803840	3
tail rotor blade (×2)	335488	3
background grid	1869824	414
auxiliary	100480	2
total	11772544	480

with 11.8 million grid cells. Grid data for the individual body components are summarized in Table 1.

5 Simulation Setup and Flow Computation

The flow simulation was set up according to the wind tunnel parameters given in Sect. 2. Since no data were available for the flapping motion of the main rotor and the coupled cyclic pitching/flapping of the tail rotor, these angles were set to zero. The elastic deformation of the blades was not taken into account. Both simplifications will introduce errors into the simulation. Future simulations will therefore use a trim procedure in order to obtain the correct blade motion.

For the flow simulation the time step was chosen to be equal to a 2° rotation of the tail rotor. This corresponds to a rotation of 0.4° of the main rotor. Therefore, a complete revolution of the main rotor requires 900 time steps. Within each physical time step, 50 iterations of the flow solver were performed in order to converge the dual-time stepping method.

The simulation was executed on the NEC SX-6 vector computer at the High Performance Computing Center in Stuttgart. One node of the machine with eight processors was used. The computation required 12 gigabytes of memory and run for four weeks. Within this time 2.3 revolutions of the main rotor were computed. This is sufficient to obtain a periodic solution, since due to the high inflow velocities any disturbances are quickly transported downstream.

During the simulation more than 400 gigabyte of data were produced. This huge amount of data posed a major issue on transfering the data to local computers, to store them and to do the postprocessing.

6 Results

In this section a brief overview of the results is given. A more detailed discussion can be found in [20]. The computed pressure distribution for the symmetry plane of the fuselage in comparison with experimental data is shown in Fig. 5. The agreement of experimental and computed data is very good. By comparing Fig. 5, left and Fig. 5, right, unsteady pressure variations can be noticed at the tail boom and the fin of the helicopter. On the nose of the helicopter, only a little effect of the unsteadiness can be seen.

The pressure distributions for four different positions of a main rotor blade are presented in Fig. 6. The overall agreement between the computed and measured data is good. At $\Psi = 180°$ some larger differences can be observed. These are due to the elastic blade deformation, which has not been taken into account during the simulation.

Figure 7 presents the distributions of the tail rotor pressure for the radial position $r/R = 0.87$. At azimuth angle $\Psi_{TR} = 0°$ the tail rotor blade points downwards. From the pressure patterns at $\Psi_{TR} = 0$ and $\Psi_{TR} = 90$ it can be

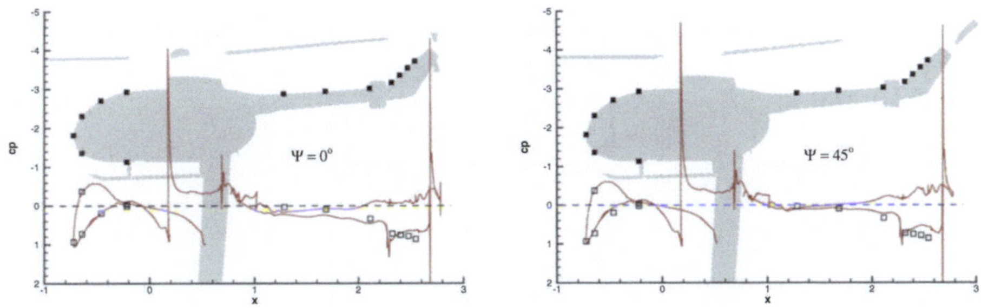

Fig. 5. Instantaneous surface pressure distribution in the symmetry plane. Comparison of computation and experiment, left $\Psi_{MR} = 0°$, right: $\Psi_{MR} = 45°$

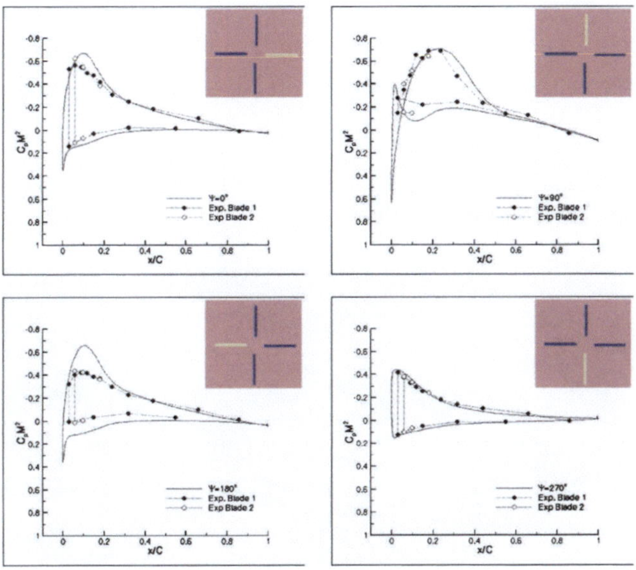

Fig. 6. Computed and measured pressure $(c_p \cdot M^2)$ on main rotor at 87% blade radius depending on main rotor azimuth angle Ψ_{MR}

deduced, that in comparison to experimental data the tail rotor in the simulation produces too much thrust on the advancing side of the rotor. The local angle of attack in the simulation is therefore higher than during the measurements. This difference can be explained by the non-consideration of the coupled cyclic pitching/flapping motion of the tail rotor blades. On the retreating side of the rotor the agreement between the measured and the computed data is good.

A snapshot of the computed vortex structure is shown in Fig. 8 in terms of constant λ_2 surfaces [21]. The figure illustrates an extremely complex flow field with several interacting vortex systems. The four blade tip vortices can clearly be seen and some blade vortex interactions can be identified. The computations also reproduce the interaction of the main rotor wake with the tail rotor. The

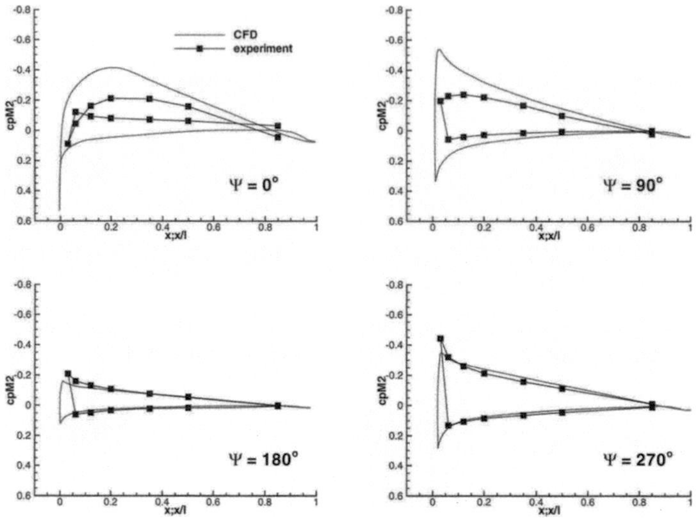

Fig. 7. Computed and measured pressure ($c_p \cdot M^2$) on tail rotor at 80% blade radius depending on tail rotor azimuth angle Ψ_{TR}

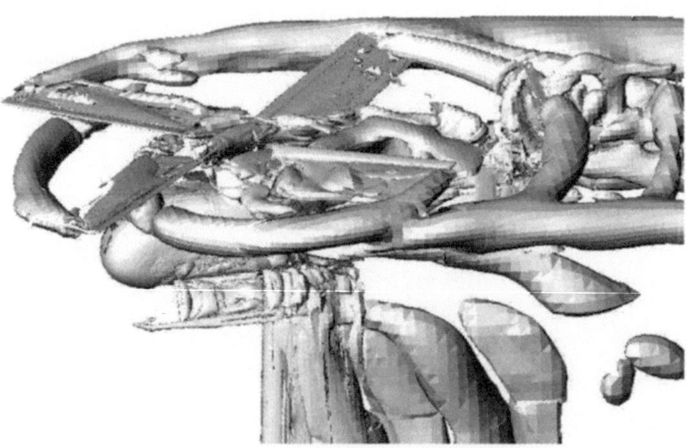

Fig. 8. Vortex cores detected with λ_2 criterion

unsteady vortex shedding from the helicopter skids and the model support can clearly be seen.

7 Computational Performance

In the past, a large effort was spent in optimizing the flow solver FLOWer for parallel computations on vector machines. Nevertheless, the total CPU time for the simulation was four weeks. This is acceptable for research purposes but much too long for industrial use in helicopter design. In this section the computational performance of FLOWer is analyzed to demonstrate some progress in efficiency and to identify parts of the code which may be subject to further improvements.

The unsteady flow simulation can be subdivided into three main parts, see the flow chart in Fig. 9. At the beginning of a new physical time step, the grids are positioned in space according to the positions of the main and tail rotor blades. In a second step, which will subsequently be called 'Chimera-part', the communication between the overlapping grids is established. To this end, holes must be cut into the grids in order to blank grid points inside solid bodies and the grids must be searched to identify donor cells for the interpolation of data. Afterwards the flow is computed for the physical time step under consideration. This is accomplished by performing 50 iterations of the flow solver in order to converge the implicit dual time stepping method.

A performance analysis was made at the beginning of the complete helicopter simulation by using eight processors of the NEC SX-6 vector computer. The study revealed that within one physical time step 460 seconds were spent in the Chimera part whereas $50 \cdot 9.3 = 465$ seconds were spent for the flow computation. The time used to position the grid can be neglected. The total execution time per time step is therefore 925 seconds, see Table 2. This shows that only half of the execution time was used to solve the flow equations while

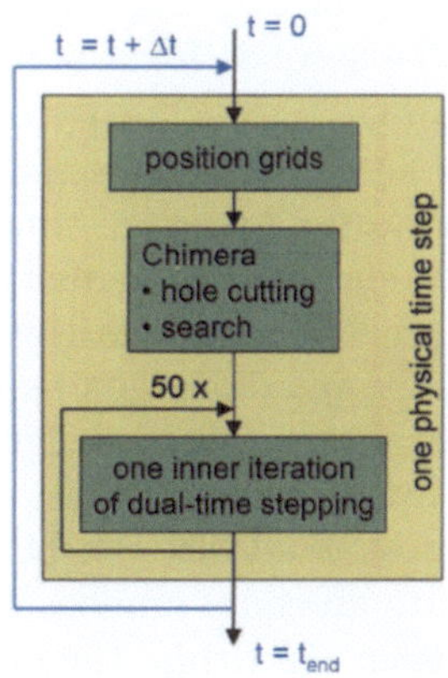

Fig. 9. Flow chart of unsteady flow simulation

Table 2. Performance improvement, parallel computations with eight processors of NEC SX-6

	Chimera	flow solver	one time step
starting point	460 s	$50 \cdot 9.3$ s	925 s
final state	69 s	$50 \cdot 9.3$ s	534 s
PC-Cluster Intel Xeon 3.06 GHz	55 s	$50 \cdot 72.6$ s	3740 s

the other time was spent in the preprocessing. In order to improve the ratio, several code modifications were made for the Chimera-part during the course of the flow simulation. The largest gain in speed was obtained by vectorizing the hole cutting procedure, which was only partly done before. Other modifications were loop unrolling and reorganization of the data flow. At the final stage, the time needed by the Chimera algorithms was 69 seconds, which is a reduction of 85% compared to the initial state, see Table 2. Therefore, one physical time step requires 534 seconds and only 13% of the total CPU-time is spent in the Chimera-part.

In order to show the efficiency of the NEC SX-6 vector computer, a performance analysis was conducted on eight processors of a PC-Cluster with Intel Xeon Processors with 3.06 GHz. The study shows that only 55 seconds are required for the Chimera part whereas $50 \cdot 72.6 = 3630$ seconds are needed by the flow solver, see Fig. 2. The time spent for the Chimera-part reveals that despite all the improvements the vector computer is slower in the Chimera-part than the scalar computer. This is due to some non-vectorized parts in the search algorithm. The outstanding performance of the vector computer becomes evident for the flow solver, which is 7.8 times faster than on the PC-Cluster.

The parallel efficiency of the FLOWer flow solver on the NEC SX-6 is presented in Table 3. While 262 seconds are required for the Chimera algorithms in sequential mode, the time is reduced to 69 seconds when using eight processors. This corresponds to a speed-up of 3.8. The time needed to converge the flow equations is reduced from 55.4 seconds for a sequential run to 9.3 seconds for a parallel run on eight processors. This is a speed-up of 6.0.

The theoretic speed-up of 8.0 for eight processors is not reached. For the flow solver part this is mainly due to the increased time needed to receive data from memory when several processes access the memory at the same time.

The reduced efficiency of the Chimera-algorithms is caused by a non-optimal load balancing. In FLOWer, the parallelization is based on domain decomposition, where the same number of grid cells is assigned to each processor. This is the optimal load balancing for the flow simulation. In order to optimally balance the Chimera-part, the grid would have to be redistributed two times: The hole cutting algorithms require an equal number of cells to be blanked on each processor, whereas for the search algorithm, the number of donor cells for interpolation must be balanced. This code improvement may become important, if in future more processors are used for simulations.

Table 3. Parallel performance on NEC SX-6

		seq	2 proc	4 proc	8 proc
Chimera	time	262.0 s	146.2 s	98.6 s	69.5 s
	speed-up		1.8	2.7	3.8
flow solver	time	55.4 s	29.3 s	16.0 s	9.25 s
(one phys. time step)	speed-up		1.9	3.5	6.0

8 Performance on NEC SX-8 Vector Computer

In spring 2005 the next generation vector computer NEC SX-8 was installed as the successor of the NEC SX-6 at the High Performance Computing Center in Stuttgart. In order to estimate the benefits for future helicopter flow computations, an evaluation of the NEC SX-8 performance will be given in this section. The parallel performance of the NEC SX-8 is presented in Table 4. The data can be directly compared to the performance of the NEC SX-6 shown in Table 3. Comparing the execution times of NEC SX-6 and NEC SX-8 for a sequential run, the Chimera part is executed 1.48 times faster on the NEC SX-8 than on the NEC SX-6, whereas the flow solver is 1.79 times faster. The differences in the improvements can be explained by the different code structure, where the Chimera-part contains many integer operations and non-vectorized if-branches, whereas the flow solution procedure has a simple code structure, is well vectorized and contains only floating point operations.

The parallel speed-up is comparable on the NEC SX-6 and on the NEC SX-8. Only when using eight processors the speed-up of the NEC SX-8 is slightly smaller than for the NEC SX-6.

Comparing the total computational time for one physical time step using eight processors, on the NEC SX-6 $69.5\text{s} + 50 \cdot 9.25\text{s} = 532\text{s}$ are required, whereas on the NEC SX-8 the wall clock time is $47.6\text{s} + 50 \cdot 5.47\text{s} = 321\text{s}$. This is equal to an overall improvement of a factor 1.66.

Table 4. Parallel performance on NEC SX-8

		seq	2 proc	4 proc	8 proc
Chimera	time	176.7 s	100.7 s	66.9 s	47.6 s
	speed-up		1.8	2.6	3.7
flow solver	time	31.0 s	16.4 s	9.03 s	5.47 s
(one phys. time step)	speed-up		1.9	3.4	5.7

9 Summary and Conclusions

A time-accurate Navier-Stokes simulation of a BO-105 helicopter wind tunnel model has been presented. The computations considered forward flight conditions at 60 m/sec. The motion of the main and tail rotor was realized numerically with the Chimera method. Periodic solutions were obtained in 650 wall clock hours using eight processors of a NEC SX-6 vector computer. Within this time 0.4 terabyte data were produced and had to be analyzed.

Very good agreement with experiment could be obtained for the fuselage. Main rotor pressure could be predicted satisfactorily. As expected, some deviation from the experimental results was observed on the advancing blade due to blade elasticity. Noticeable differences between the CFD results and experiment were found on the tail rotor. This was due to significant deviation in the rotor's pitch and flap angles from the nominal trim law which was used in the computations.

In an effort to reduce the execution time of the flow solver, the Chimera-routines were optimized and vectorized. This reduced the execution time for one physical time step by a factor of 1.7. While the presented simulation was run on a NEC SX-6, the new installation of a NEC SX-8 has speed-up the calculation by a factor of 1.7.

The reported efforts are an important step towards the simulation of helicopters with an even more detailed geometry. Future applications will include engine intake and exhaust, rotors with hubs and the elastic deformation of the blades. Furthermore, the simulation will be embedded into a trim loop by coupling a flight mechanics code to the flow solver. This type of simulation will again increase the time requirements for the flow simulations by a factor of five or even more. Further code improvements and the access to high performance platforms are therefore mandatory.

References

1. Gleize, V., Costes, M., Geyr, H., Kroll, N., Renzoni, P., Amato, M., Kokkalis, A., Muttura, L., Serr, C., Larrey, E., Filippone, A., Fischer, A.: Helicopter Fuselage Drag Prediction: State of the Art in Europe. AIAA-Paper 2001-0999, 2001
2. Beaumier, P., Chelli, E., Pahlke, K.: Navier-Stokes Prediction of Helicopter Rotor Performance in Hover Including Aero-Elastic Effects. American Helicopter Society 56th Annual Forum, Virginia Beach, Virginia, May 2–4, 2000
3. Pomin, H., Wagner, S.: Aeroelastic Analysis of Helicopter Rotor Blades on Deformable Chimera Grids. AIAA Paper 2002-0951
4. Pahlke, K., van der Wall, B.: Chimera Simulations of Multibladed Rotors in High-Speed Forward Flight with Weak Fluid-Structure-Interaction. Aerospace Science and Technology, Vol 9. pp. 379–389, 2005
5. Le Chuiton, F.: Actuator Disc Modelling For Helicopter Rotors. Aerospace Science and Technology, Vol. 8, No. 4, pp. 285–297, 2004
6. Meakin, R. B.: Moving Body Overset Grid Methods for Complete Aircraft Tiltrotor Simulations. AIAA-Paper 93-3359, 1993
7. Khier, W., le Chuiton, F., Schwarz, T.: Navier-Stokes Analysis of the Helicopter Rotor-Fuselage Interference in Forward Flight. CEAS Aerospace Aerodynamics Research Conference, Cambridge, England, June 10-12, 2002
8. Renauld, T., Le Pape, A., Benoit, C.: Unsteady Euler and Navier-Stokes computations of a complete helicopter. 31st European Rotorcraft Forum. Florence, Italy, September 13–15, 2005
9. Sides, J., Pahlke, K., Costes, M.: Numerical Simulation of Flows Around Helicopters at DLR and ONERA. Aerospace Science and Technology, Vol. 5, pp 35–53, 2001
10. Pahlke, K., Costes, M., D'Alascio, A., Castellin, C., Altmikus, A.: Overview of Results Obtained During the 6-Year French-German Chance Project. 31st European Rotorcraft Forum, Florence, Italy, September 13–15, 2005

11. Langer, H.-J., Dieterich, O., Oerlemans, S., Schneider, O., van der Wall, B., Yin, J.: The EU HeliNOVI Project – Wind Tunnel Investigations for Noise and Vibration Reduction. 31st European Rotorcraft Forum, Florence, Italy, September 13–15, 2005
12. Yin, J., van der Wall, B., Oerlemans S.: Representative Test results from HeliNOVI Aeroacoustic Main Rotor/Tail Rotor/Fuselage Test in DNW. 31st European Rotorcraft Forum, Florence, Italy, September 13–15, 2005
13. Jameson, A., Schmidt, W., Turkel, E.: Numerical Solutions of the Euler Equations by Finite Volume Methods using Runge-Kutta Time-Stepping Schemes. AIAA-Paper 81-1259, 1981
14. Wilcox, D. C.: Reassessment of the Scale-Determining Equation for Advanced Turbulence Models. AIAA Journal, vol. 26, no. 11, November 1988
15. Rudnik, R.: Untersuchung der Leistungsfähigkeit von Zweigleichungs-Turbulenzmodellen bei Profilumströmungen. Deutsches Zentrum für Luft- und Raumfahrt e.V., FB 97-49, 1997
16. Jameson, A.: Time Dependent Calculations Using Multigrid, with Applications to Unsteady Flows Past Airfoils and Wings. AIAA-paper 91-1596, 1991
17. Melson, N. D., Sanetrik, M. D., Atkins, H. L.: Time-Accurate Navier-Stokes Calculations with Multigrid Acceleration. Proceedings of the 6th Copper Mountain Conference on Multigrid Methods, NASA Conference Publication 3224, 1993, pp. 423–439
18. Benek, J. A., Steger, J. L., Dougherty, F. C.: A Flexible Grid Embedding Technique with Application to the Euler Equations. AIAA-Paper 83-1944, 1983
19. Schwarz, T.: The Overlapping Grid Technique for the Time-accurate Simulation of Rotorcraft Flows. 31st European Rotorcraft Forum, Florence, Italy, September 13–15, 2005
20. Khier, W., Schwarz, T., Raddatz, J.: Time-accurate Simulation of the Flow around the Complete BO-105 Wind Tunnel Model. 31st European Rotorcraft Forum, Florence, Italy, September 13–15, 2005
21. Jeong, J., Hussain, F.: On the identification of a vortex. Journal of Fluid Mechanics, vol. 285, pp. 69–94, 1995

A Hybrid LES/CAA Method for Aeroacoustic Applications

Qinyin Zhang, Phong Bui, Wageeh A. El-Askary, Matthias Meinke, and Wolfgang Schröder

Institute of Aerodynamics, RWTH Aachen,
Wüllnerstrasse zwischen 5 und 7,
D-52062 Aachen, Germany,
office@aia.rwth-aachen.de

Abstract This paper describes a hybrid LES/CAA approach for the numerical prediction of airframe and combustion noise. In the hybrid method first a Large-Eddy Simulation (LES) of the flow field containing the acoustic source region is carried out from which then the acoustic sources are extracted. These are then used in the second computational Aeroacoustics (CAA) step in which the acoustic field is determined by solving linear acoustic perturbation equations. For the application of the CAA method to a unconfined turbulent flame, an extension of the method to reacting flow fields is presented. The LES method is applied to a turbulent flow over an airfoil with a deflected flap at a Reynolds number of $Re = 10^6$. The comparison of the numerical results with the experimental data shows a good agreement which shows that the main characteristics of the flow field are well resolved by the LES. However, it is also shown that a zonal LES which concentrates of the trailing edge region on a refined local mesh leads to a further improvement of the accuracy. In the second part of the paper, the CAA method with the extension to reacting flows is explained by an application to a non-premixed turbulent flame. The monopole nature of the combustion noise is clearly verified, which demonstrates the capability of the hybrid LES/CAA method for noise prediction in reacting flows.

1 Introduction

In aeroacoustics turbulence is often a source of sound. The direct approach of noise computation via a Direct Numerical Simulation (DNS) including all turbulent scales without any modelling is still restricted to low-Reynolds number flows and for real technical applications computationally too expensive. An attractive alternative is the Large Eddy Simulation (LES) which resolves only the turbulent scales larger than the cell size of the mesh. In order to take advantage of the disparity between fluid mechanical and acoustical length scales it is reasonable to separate the noise computation into two parts. In a first step, the LES resolves the acoustic source region governed by nonlinear effects. In a second step the acoustic field is computed on a coarser grid by linear acoustic equations with the nonlinear effects lumped together in a source term being calculated from the LES. The feasibility of such a hybrid LES/CAA method is demonstrated in the following by means of two different applications, airframe noise and combustion noise. The first part of the two-step method, the LES computation, is shown for airfoil flow. The present results will comprise a detailed analysis of the turbulent scales upstream of the trailing-edge, a thorough investigation of the surface pressure fluctuations and the trailing-edge eddies generated in the near-field wake based on LES findings. The simulation provides the data that allows the acoustic source functions of the acoustic wave equations to be evaluated. The second part, the application of the CAA method, is performed for combustion noise of unconfined turbulent flames. The acoustic analogy used for the acoustic field computation is the system of the Acoustic Perturbation Equations (APE) which have been extended to take into account noise generation by reacting flow effects.

2 LES for Trailing Edge Noise

Turbulent flow near the trailing edge of a lifting surface generates intense, broadband scattering noise as well as surface pressure fluctuations. The accuracy of the trailing-edge noise prediction depends on the prediction method of the noise-generating eddies over a wide range of length scales. Recent studies indicate that promising results can be obtained when the unsteady turbulent flow fields are computed via large eddy simulation (LES). The LES data can be used to determine acoustic source functions that occur in the acoustic wave propagation equations which have to be solved to predict the aero-acoustical field [1].

The turbulence embedded within a boundary layer is known to radiate quadrupole noise which in turn is scattered at the trailing edge [2]. The latter gives rise to an intense noise radiation which is called trailing-edge noise. Turbulence is an insufficient radiator of sound, particularly at low Mach number, $M_\infty \leq 0.3$, meaning that only a relatively small amount of energy is radiated away as sound from the region of the turbulent flow. However, if the turbulence interacts with a trailing edge in the flow, scattering occurs which changes the inefficient quadrupole radiation into a much more efficient dipole radiation [2].

The noise generated by an airfoil is attributed to the instability of the upper and lower surface boundary layers and their interactions with the trailing edge [3]. The edge is usually a source of high-frequency sound associated with smaller-scale components of the boundary layer turbulence. Low-frequency contributions from a trailing edge, that may in practice be related to large-scale vortical structures shed from an upstream perturbation, are small because the upwash velocity they produce in the neighborhood of the edge tends to be canceled by that produced by vorticity shed from the edge [3]. For this purpose, a large-eddy simulation is carried out to simulate the three-dimensional compressible turbulent boundary layer past an airfoil-flap configuration and airfoil trailing edge which can be used for an acoustic simulation.

3 Computational Setup

One of the primary conclusions from the European LESFOIL project is that an adequate numerical resolution especially in the near wall region is required for a successful LES. On meshes which do not resolve the viscous near-wall effects, neither SGS models nor wall models were able to remedy these deficiencies [4]. According to the experience from LES of wall bounded flows, the resolution requirements of a wall resolved LES are in the range of $\Delta x^+ \approx 100$, $\Delta y^+ \approx 2$ und $\Delta z^+ \approx 20$, where x, y, z denote the streamwise, normal, and spanwise coordinates, respectively. During the mesh generation process, it became clear that it would lead to meshes with unmanageable total grid point numbers, if these requirements are to be followed strictly. For the preliminary study, the streamwise resolution is approximately $\Delta x^+ \approx 200 \sim 300$, whereas the resolution is set to $\Delta y^+_{min} \approx 2$ and $\Delta z^+ \approx 20 \sim 25$ in the wall normal and in the spanwise direction, respectively. The mesh of the preliminary study is shown in Fig. 1. The extent of the computational domain is listed in Table 1.

The spanwise extent of the airfoil is 0.32 percent of the chord length. A periodic boundary condition is used for this direction. The relatively small spanwise extent is chosen because of the following two reasons. On the one hand, it was shown for flat plate flows that due to the high Reynolds number of $Re = 1.0 \times 10^6$, two-point correlations decay to zero already in about 250 wall units [5] and on the other hand, since a highly resolved mesh was used in the wall normal and tangential direction the computational domain must be limited to a reasonable size to reduce the overall computational effort. Towards the separation regions in the flap cove and at the flap trailing edge, the boundary layer thicknesses and the characteristic sizes of turbulent structures will increase. Therefore, especially

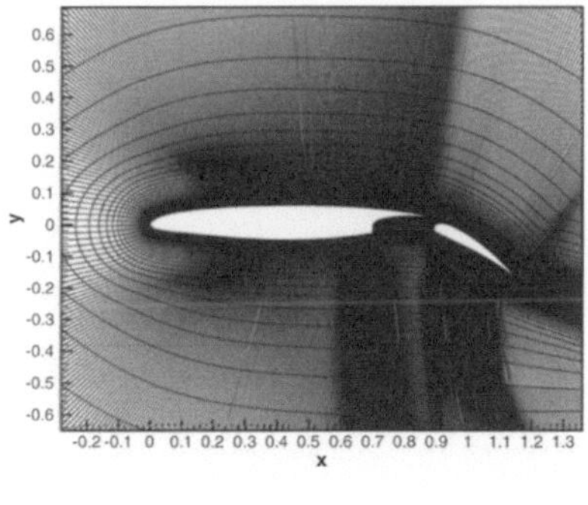

Fig. 1. Computational mesh in the x–y plane, which shows every second grid point

Table 1. Computational domain and grid point distribution, SWING+ Airfoil ($Re = 10^6$)

L_X	L_Y	L_Z	$N_x \times N_y$	N_z	Total Grid Points
−4.0 ... 5.0	−4.0 ... 4.0	0 ... 0.0032	241.320	17	4.102.440

for these areas, the spanwise extent of the computational domain needs to be increased in further computations. A no-slip boundary condition is applied at the airfoil surface, and a non-reflecting boundary condition according to Poinsot & Lele [6] is used for the far field.

Another possibility to enlarge the spanwise extent without increasing the total number of grid points and as such the overall computational effort, is to focus the analysis on the zone of the flow field that is of major interest for the sound propagation. For the airfoil flow problem this means that a local analysis of the flow in the vicinity of the trailing edge has to be pursued. To do so, we apply the rescaling method by El-Askary [5] which is valid not only in compressible flows but also in flows with weak pressure gradients. This procedure is a consistent continuation of the approach that was already successfully applied in the analysis of the trailing edge flow of a sharp flat plate [1, 7, 8, 9, 10].

4 Results

In the following, we discuss first the LES of the airfoil-flap configuration and then turn to the zonal analysis of the airfoil trailing edge flow. In both problems numerical and experimental data are juxtaposed.

In the flow over the SWING+ airfoil with deflected flap, separations exist in the flap cove and at the flap trailing edge. In the current numerical simulation, both separation regions are well resolved. They are visible in the time and spanwise averaged streamlines of the numerical simulation (Fig. 2).

The turbulent flow is considered to be fully developed such that the numerical flow data can be time averaged over a time period of $\Delta T = 3.0\ c/u_{\infty}$, and then spanwise averaged to obtain mean values. The distribution of the mean pressure coefficient c_p over the airfoil surface is plotted in Fig. 4. A good agreement with the experimental data [11] is achieved. It is worth mentioning, that the onset of the turbulent flow separation and the size of the separation bubble in the vicinity of the flap trailing edge are captured quite exactly by the numerical simulation. This can be seen by the plateau in the c_p-distribution near the trailing edge of the flap.

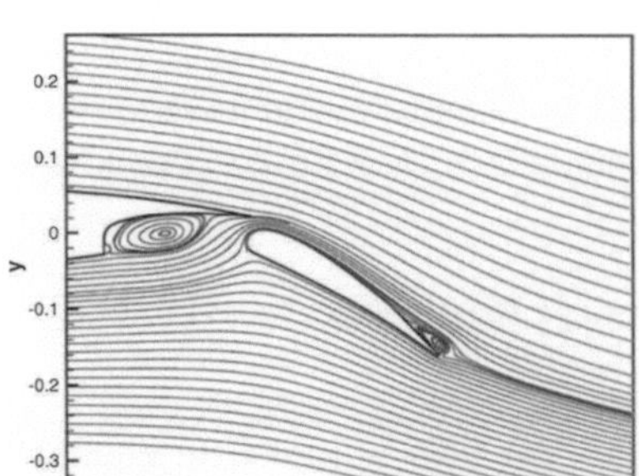

Fig. 2. Time and spanwise averaged streamlines

In the experiments carried out by the Institute of Aerodynamics and Gasdynamics of the University Stuttgart, profiles of the wall tangential velocity were measured at five locations two of which were on the airfoil and three on the flap [11]. The mean velocity profiles from the experiments and the numerical simulation are compared in Fig. 5. The qualitative trends of the velocity profiles agree with each other. Note that in the near wall region of position D, the measuring location lies in the separation area, where the hot wire probe cannot obtain the correct velocity information. This is the reason for the different signs of the velocity profiles in the near wall region.

As can be expected from the good agreement of the c_p distribution over the airfoil, the lift coefficient agrees very well with the experimental value [11]. The drag coefficient, however, is over-predicted by the numerical simulation, which is not surprising due to the discrepancies in the velocity profiles.

We now turn to the discussion of the zonal large eddy simulation of an airfoil trailing-edge flow at an angle of attack of $\alpha = 3.3$ deg. The trailing-edge length constitutes 30% of the chord length c. The computational domain is shown in Fig. 6.

All flow parameters are given in Table 2, in which Re_c is the Reynolds number based on the chord length. In the present simulation the inflow section is located at a position of an equilibrium turbulent boundary layer with a weak adverse

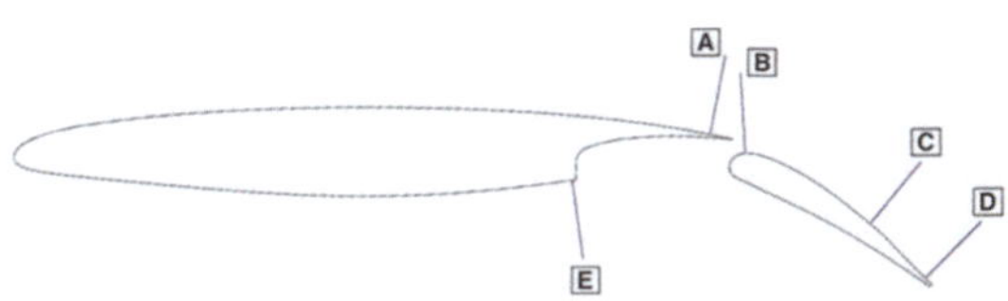

Position	A	B	C	D	E
x/c	0.8675	0.9100	1.0675	1.1375	0.6950

Fig. 3. Locations of the velocity measurements on the airfoil and flap

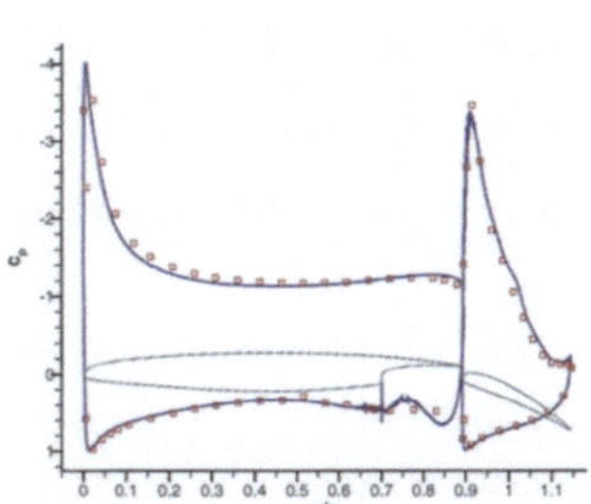

Fig. 4. Time and spanwise averaged pressure coefficients. Solid line: numerical data, Symbols: experimental data [11]

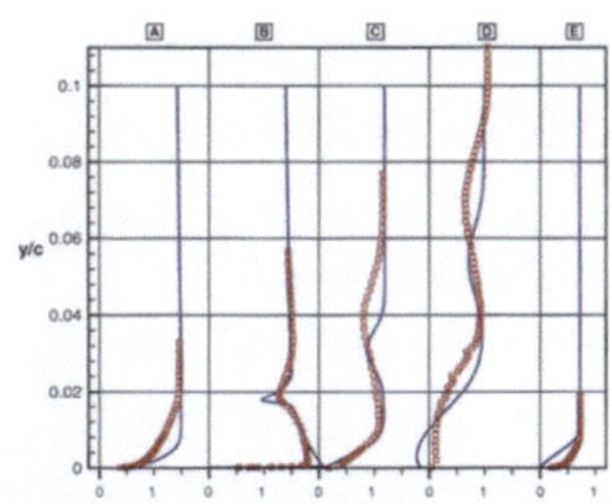

Fig. 5. Time and spanwise averaged profiles of the wall tangential velocity components. Solid line: numerical data, Symbols: experimental data [11]

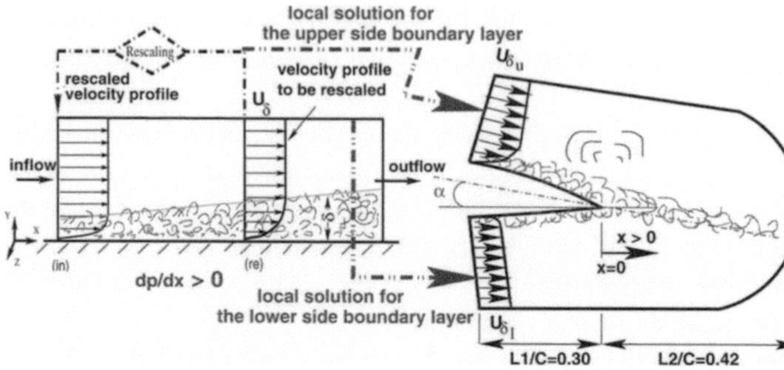

Fig. 6. Computational domain for the airfoil trailing edge flow (right) and inflow distribution from a flat plate boundary layer with an equilibrium adverse pressure gradient via the slicing technique

pressure gradient. Therefore, all inflow data for the upper side boundary layer of the airfoil δ_u ($\delta_u = 0.01972c$) and for the lower side boundary layer δ_l ($\delta_l = 0.0094745c$) are extracted from two separate LES of flat plate boundary layers with an equilibrium pressure gradient, which use the new rescaling formulation for a variable pressure pressure gradient. A total of 8.9×10^6 computational cells are employed with mesh refinements near the surface and the trailing edge (Fig. 7). The resolution used for the present results is $\Delta y^+_{\mathrm{min}} \approx 2$, $\Delta z^+ \approx 32$, and $\Delta x^+ \approx 87$ at the inlet and $\Delta x^+ \approx 5$ near the trailing edge, see Table 2.

The vortex structures in the boundary layer near the trailing edge and in the near wake are presented by the λ_2 contours in Fig. 8. A complex structure can be observed immediately downstream of the trailing edge. This is due to the interaction of two shear layers shedding from the upper and lower airfoil surface. An instantaneous streamwise velocity field in the mid-span is plotted in Fig. 9. Note that in the velocity distribution a small recirculation region occurs right downstream of the trailing edge.

Comparisons of the mean streamwise velocity profiles with experimental data of [11] are presented in Figs. 10(a) and 10(b) for the upper and lower side, respectively, at several streamwise locations: $x/c = -0.1, -0.05, -0.02, 0.0$ measured from the trailing edge. Whereas in the near wall region a good agreement is ob-

Table 2. Parameters and domain of integration for the profile trailing-edge flow simulation. See also Fig. 6

Re_c	$Re_{\delta o}$	δ_o/c	M_∞
$8.1 \cdot 10^5$	15989	0.01972	0.15
L_1	L_2	L_z	grid points
$0.3\ c$	$0.42\ c$	$0.0256\ c$	$8.9 \cdot 10^6$
$\Delta x^+_{\mathrm{min}}$	$\Delta x^+_{\mathrm{max}}$	$\Delta y^+_{\mathrm{min}}$	Δz^+
5	87	2	32

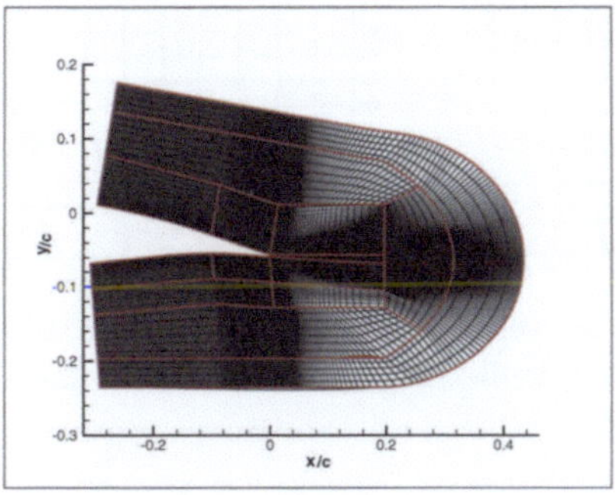

Fig. 7. LES grid of a turbulent flow past an airfoil trailing edge

served between the computed and measured mean-velocity profiles. Pronounced deviations occur in the log- and outer region of the boundary layer, which could be caused by the coarsening of the mesh.

Further downstream of the trailing edge, asymmetric wake profiles are observed as shown in Fig. 10(c) in comparison with the experimental data. This asymmetry is generated by varying shear layers on the upper and lower surface shed from the trailing edge. Even at $x/c = 0.145$ no fully symmetric velocity distribution is regained. Note the good qualitative and quantitative experimental and numerical agreement for the velocity distributions.

For the analysis of airfoil flow the skin-friction coefficient is one of the most critical parameters. Its distribution evidences whether or not the flow undergoes separation. Comparisons of the present computations with the experimental values are shown in Figs. 11(a) and 11(b) on the upper and lower surface, respectively. The simulation results are in good agreement with the data of [11] except right at the end of the trailing edge. This could be due to an insufficient numerical resolution near the trailing edge or could be caused by some inaccuracies in the experimental data in this extremely susceptible flow region.

The simulations were carried out on the NEC SX-5 and NEC SX-6 of the High Performance Computing Center Stuttgart (HLRS). The vectorization rate of the

Fig. 8. Vortex structures in the boundary layer near the trailing edge and in the near wake (λ_2 contours)

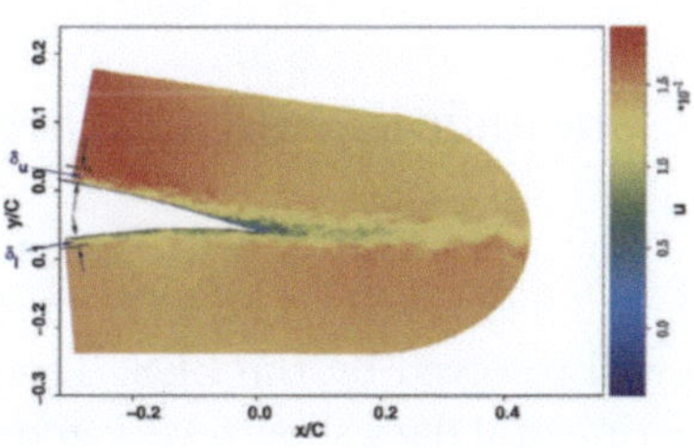

Fig. 9. Instantaneous velocity contours in the boundary layer near the trailing edge and in the near wake

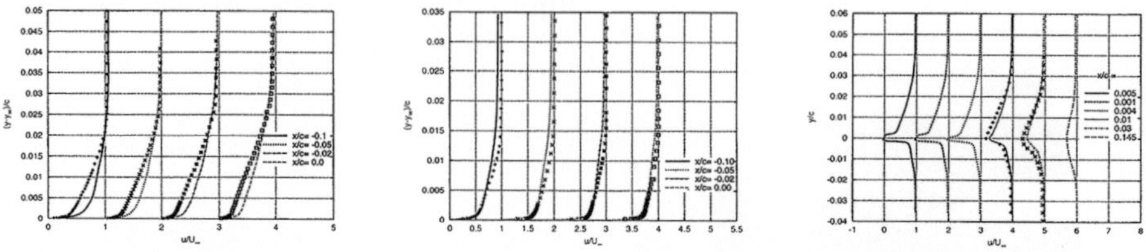

(a) Mean streamwise velocity profiles near the trailing edge (upper side).

(b) Mean streamwise velocity profiles near the trailing edge (lower side).

(c) Mean streamwise velocity profiles in the wake.

Fig. 10. Mean streamwise velocity profiles compared with experimental data (symbols) [11]

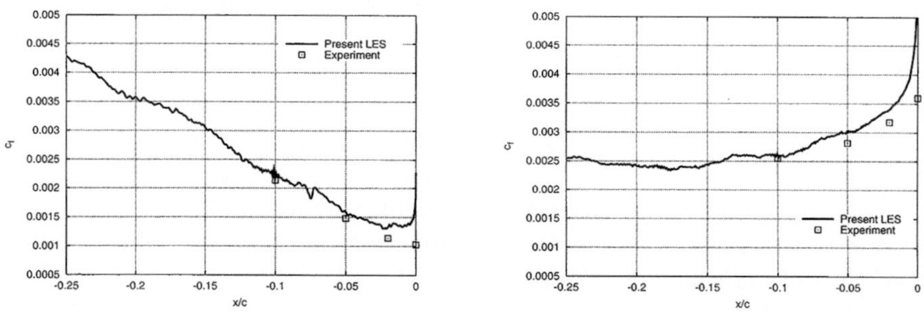

(a) Skin-friction coefficient on the upper side of the trailing edge.

(b) Skin-friction coefficient on the lower side of the trailing edge.

Fig. 11. Skin-friction coefficients compared with experimental data (symbols) [11]

flow solver is 99%, and a single processor performance of about 2.4 GFlops on a SX-5 and 4.3 GFlops on a SX-6 processor is achieved. The memory requirement for the current simulation is around 3.5 GB. Approximately 175 CPU hours on 10 SX-5 CPUs for statistically converged solution data are required for the airfoil-flap configuration and roughly 75 CPU hours on 10 SX-5 processors for the zonal approach.

5 Conclusions and Outlook for Airfoil Flow

The flow over an airfoil with deflected flap at a Reynolds number of $Re = 10^6$ has been studied based on an LES method. The main characteristics of the flow field are well resolved by the LES. The comparison of the numerical results with the experimental data show a very good match of the pressure coefficient distribution and a qualitative agreement of the velocity profiles. The results achieved to date are preliminary but encouraging for further studies. The main reason for the deficiency in the numerical results is the fact, that the resolution requirement for an LES cannot be met everywhere in the computational domain at this high Reynolds number.

The zonal approach results in a pronounced improvement of the local accuracy of the solution. The skin-friction coefficient distribution and the near wall as well as the wake velocity profiles show a convincing agreement with the experimental data.

The experience with the present global LES method evidences, that good results can be achieved if the resolution requirements are met [5]. For this reason, the next step will be to concentrate on the improvement of the computational setup. Since the outer part of the flow field over an airfoil is predominantly two-dimensional and laminar, only a quasi-2d calculation will be performed in this area in the next step. For this purpose, a 2D/3D coupling technique has been developed for the structured solver. With this technique, it is possible to increase the near wall resolution while keeping the overall computational cost at a relatively low level. Next, hybrid RANS/LES coupling techniques are contemplated for the improvement of the overall numerical method. Furthermore, with respect to the simulation of the sound field the LES data from the zonal approach will be postprocessed to determine the source terms of the acoustic perturbation equations, which were already successfully used in [1].

6 CAA for Combustion Noise

This research project is part of the Research Unit FOR 486 “Combustion Noise”, which is supported by the German Research Council (DFG). The objective of the Institute of Aerodynamics of the RWTH Aachen University is to investigate the origin of combustion noise and its mechanisms. The LES for the two-step approach is performed by the Institute for Energy and Powerplant Technology from Darmstadt University of Technology, followed by the CAA simulation to compute the acoustical field. This hybrid LES/CAA approach is similar to that in [1]. However, in this study the Acoustic Perturbation Equations are extended to reacting flows. In flows, where chemical reactions have to be considered, the application of such an approach is essential as the disparity of the characteristic fluid mechanical and acoustical length scales is even more pronounced than in the non-reacting case.

It is well known from the literature, e.g. [12, 13], that noise generated by combustion in low Mach number flows is dominated by heat release effects, whereas in jet or airframe noise problems the major noise contribution originates from the Lamb vector ($\boldsymbol{L}' = (\boldsymbol{\omega} \times \boldsymbol{u})'$), which can be interpreted as a vortex force [14, 15]. In principle it is possible to treat this task by extending Lighthill's Acoustic Analogy to reacting flows as was done in the past [12, 13]. This, however, leads to an inhomogeneous wave equation with an ordinary wave operator e.g. [13, 16], which is valid for homogeneous mean flow only. Therefore, this approach is restricted to the acoustic far field. The APE approach remedies this drawback. It is valid in non-uniform mean flow and takes into account convection and refraction effects, unlike the linearized Euler equations [14].

7 Governing Equations

To derive the extended APE system the governing equations of mass, momentum, and energy for reacting flows are rearranged such that the left-hand side describes the APE-1 system [14], whereas the right-hand side (RHS) consists of all non-linear flow effects including the sources related to chemical reactions.

$$\frac{\partial \rho'}{\partial t} + \nabla \cdot (\rho' \bar{\boldsymbol{u}} + \bar{\rho} \boldsymbol{u}') = q_c \tag{1}$$

$$\frac{\partial \boldsymbol{u}'}{\partial t} + \nabla (\bar{\boldsymbol{u}} \cdot \boldsymbol{u}') + \nabla \left(\frac{p'}{\bar{\rho}} \right) = \boldsymbol{q_m} \tag{2}$$

$$\frac{\partial p'}{\partial t} - \bar{c}^2 \frac{\partial \rho'}{\partial t} = q_e \tag{3}$$

As was mentioned before, the heat release effect dominates the generation of combustion noise. Therefore the investigations have been performed using q_e only, i.e. assuming $q_c = 0$ and $\boldsymbol{q_m} = 0$.

7.1 Thermoacoustic Source Terms

In the proposed APE system the source term containing heat release effects appears on the RHS of the pressure-density relation, i.e. q_e. This term vanishes when only isentropic flow is considered. However, due to the unsteady heat release in a flame the isentropic pressure-density relation is no longer valid in the combustion area. Nevertheless, it is this effect, which defines the major source term in comparison to the sources (q_c, $\boldsymbol{q_m}$) in the mass and momentum equations within the APE system. Concerning the other source mechanisms, which lead to an acoustic multipole behavior though it can be conjectured that they are of minor importance in the far field. Using the energy equation for reacting flows the pressure-density relation becomes:

$$\begin{aligned}
\frac{\partial p'}{\partial t} - \bar{c}^2 \frac{\partial \rho'}{\partial t} &= -\bar{c}^2 \cdot \frac{\partial \rho_e}{\partial t} \\
&= \bar{c}^2 \left[\frac{\bar{\rho}}{\rho} \cdot \frac{\alpha}{c_p} \cdot \left(\sum_{n=1}^{N} \left. \frac{\partial h}{\partial Y_n} \right|_{\rho, p, Y_m} \rho \frac{DY_n}{Dt} + \nabla \cdot \boldsymbol{q} - \frac{\partial u_i}{\partial x_j} \tau_{ij} \right) \right. \\
&\quad - \nabla (\boldsymbol{u} \rho_e) - \frac{1}{\bar{c}^2} \left[\left(1 - \frac{\bar{\rho} \bar{c}^2}{\rho c^2} \right) \cdot \frac{Dp}{Dt} - \frac{p - \bar{p}}{\rho} \cdot \frac{D\rho}{Dt} \right] \\
&\quad \left. + \left[-\frac{\gamma - 1}{\gamma} \boldsymbol{u} \cdot \nabla \bar{p} - \frac{p}{\bar{c}^2} \cdot \boldsymbol{u} \left(\frac{\nabla \bar{p}}{\bar{p}} - \frac{\nabla \bar{\rho}}{\bar{\rho}} \right) \right] \right]
\end{aligned} \tag{4}$$

where ρ_e is defined as

$$\rho_e = (\rho - \bar{\rho}) - \frac{p - \bar{p}}{\bar{c}^2} \tag{5}$$

Perturbation and time averaged quantities are denoted by a prime and a bar, respectively. The volumetric expansion coefficient is given by α and c_p is the specific heat capacity at constant pressure. For an ideal gas the equation $\alpha/c_p = (\gamma-1)/c^2$ holds. The quantity Y_n is the mass fraction of the nth species, h the enthalpy and $\boldsymbol{q}$ the heat flux.

7.2 Evaluation of the Thermoacoustic Source Terms

The investigations have been performed by considering q_e only. Reformulating the energy equation for a gas with N species [13] leads to

$$\frac{D\rho}{Dt} = \left[\frac{1}{c^2}\frac{Dp}{Dt} + \frac{\alpha}{c_p}\cdot\left(\sum_{n=1}^{N}\left.\frac{\partial h}{\partial Y_n}\right|_{\rho,p,Y_m}\rho\frac{DY_n}{Dt} + \nabla\cdot\boldsymbol{q} - \frac{\partial u_i}{\partial x_j}\tau_{ij}\right)\right] \tag{6}$$

Since the combustion takes place at ambient pressure and the pressure variations due to hydrodynamic flow effects are of low order, the whole combustion process can be assumed to be at constant pressure. From our analysis [15] and from literature [13] it is known that combustion noise is dominated by heat release effects and that all other source mechanisms are of minor importance. Assuming combustion at constant pressure and neglecting all mean flow effects q_e reduces to sources, which are related to heat release effects, non-isomolar combustion, heat flux and viscous effects. Adding up all these sources under the aforementioned restrictions the RHS of the pressure-density relation can be substituted by the total time derivative of the density multiplied by the square of the mean speed of sound and the ratio of the mean density and the density

$$q_e = \bar{c}^2\left[\frac{\bar{\rho}}{\rho}\cdot\frac{\alpha}{c_p}\cdot\left(\sum_{n=1}^{N}\left.\frac{\partial h}{\partial Y_n}\right|_{\rho,p,Y_m}\rho\frac{DY_n}{Dt} + \nabla\cdot\boldsymbol{q} - \frac{\partial u_i}{\partial x_j}\tau_{ij}\right)\right] \tag{7}$$

$$= \bar{c}^2\frac{\bar{\rho}}{\rho}\frac{D\rho}{Dt}. \tag{8}$$

8 Numerical Method

8.1 LES of the Turbulent Non-Premixed Jet Flame

In the case of non-premixed combustion, the chemical reactions are limited by the physical process of the mixing between fuel and oxidizer. Therefore, the flame is described by the classical mixture fraction approach by means of the conserved scalar f. The filtered transport equations for LES are solved on a staggered cylindrical grid of approximately 10^6 cells by *FLOWSI*, an incompressible finite-volume solver. A steady flamelet model in combination with a presumed β-Pdf approach is used to model the turbulence chemistry interaction. The subgrid stresses are closed by a Smagorinsky model with a dynamic procedure by Germano [17]. For the spatial discretization, a combination of second-order central differencing and total-variation diminishing schemes is applied [18]. The time integration is performed by an explicit third-order, low storage Runge-Kutta scheme. At the nozzle exit, time averaged turbulent pipe flow profiles are superimposed with artificially generated turbulent fluctuations [19], while the coflow is laminar.

8.2 Source Term Evaluation

The total time derivative of the density, which defines the major source term of the APE system, has been computed by the unsteady flow field in a flame region where the main heat release occurs (Fig. 12).

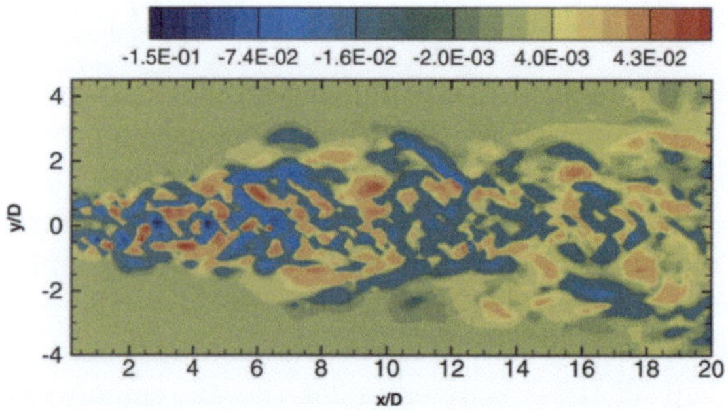

Fig. 12. Contours of the total time derivative of the density ($D\rho/Dt$) at $t = 100$ in the streamwise center plane

8.3 Grid Interpolation

Since the source terms have been calculated on the LES grid they need to be interpolated on the CAA grid. Outside the source area the APE system becomes homogeneous. This means, the RHS is defined in the source region only. Therefore, the CAA domain has been decomposed into a multiblock domain such that one block contains the entire source area. This procedure possesses the advantages that the interpolation from the LES grid to the CAA source block is much faster than onto the whole CAA domain and that the resulting data size for the CAA computation can be reduced dramatically. The data interpolation is done with a trilinear algorithm.

8.4 CAA Computation

For the CAA computation this proposed APE-System has been implemented into the PIANO (Perturbation Investigation of Aeroacoustic Noise) Code from the DLR (Deutsches Zentrum für Luft- und Raumfahrt e.V.).
The source terms on the right-hand side of the APE system has to be interpolated in time during the CAA computation. Using a quadratic interpolation method at least 25 points per period are required to achieve a sufficiently accurate distribution. Hence, the maximal resolvable frequency is $f_{max} = 1/(25\Delta t) = 800\text{Hz}$ since the LES solution comes with a time increment of $\Delta t = 5 \cdot 10^{-5}\text{s}$ [20]. This

frequency is much smaller than the Nyquist frequency. The CAA code is based on the fourth-order DRP scheme of Tam and Webb [21] for the spatial discretization and the alternating LDDRK-5/6 Runge-Kutta scheme for the temporal integration [22]. At the far field boundaries a sponge-layer technique is used to avoid unphysical reflections into the computational domain.

Solving the APE system means to solve five equations (3D) for the perturbation quantities ρ', u', v', w' and p' per grid point and time level. No extra equations for viscous terms and chemical reaction need to be considered since these terms can be found on the RHS of the APE system and are provided by the LES within the source region. On the other hand the time step within the CAA computation can be chosen much higher than in the LES. This means, using a rough estimation, that the ratio of the computation times between LES and CAA is approximately $t_{LES}/t_{CAA} \approx 4/1$.

9 Results

Figure 13 shows a snapshot of the acoustic pressure field in the streamwise center plane at the dimensionless time $t = 100$. The source region is evidenced by the dashed box. This computation was done on a 27-block domain using approximately 4×10^6 grid points, where the arrangement of the blocks is arbitrary provided that one block contains all acoustical sources.

The acoustic directivity patterns (Fig. 14) are computed for different frequencies on a circle in the $z = 0$ plane with a radius $R/D = 17$ whose center point is at $\boldsymbol{x} = (10, 0, 0)$. The jet exit diameter is denoted by D. From 150° to 210° the directivity data is not available since this part of the circle is outside of the computational domain. In general an acoustic monopole behaviour with a small directivity can be observed since this circle is placed in the acoustic near field.

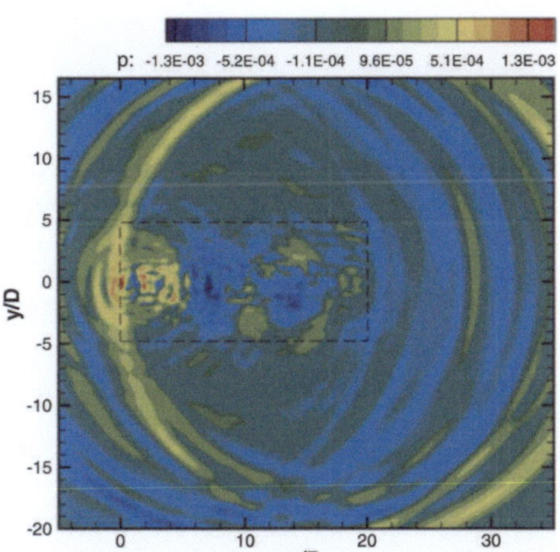

Fig. 13. Pressure contours of the APE solution at $t = 100$ in the streamwise center plane

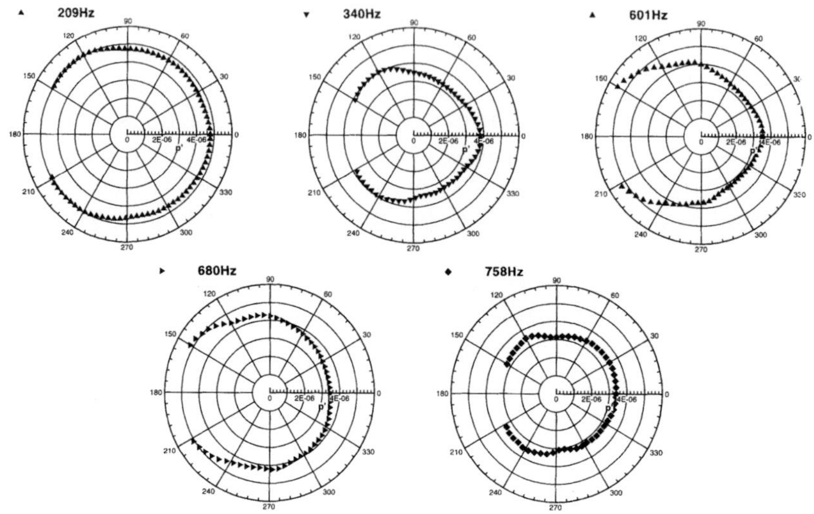

Fig. 14. Directivity patterns for different frequencies

10 Conclusion

The APE system has been extended to compute noise generated by reacting flow effects. The heat release per unit volume, which is expressed in the total time derivative of the density, represents the major source term in the APE system when combustion noise is analyzed. The main combustion noise characteristic, i.e., the monopole nature caused by the unsteady heat release, could be verified. In the present work we have demonstrated that the extended APE System in conjunction with a hybrid LES/CAA approach and with the assumptions made, is capable of simulating an acoustic field of a reacting flow, i.e., of a non-premixed turbulent flame.

Acknowledgements

The authors would like to thank the Institute for Energy and Powerplant Technology from Darmstadt University of Technology for providing the LES data of the non-premixed flame.

References

1. Ewert, R., Schröder, W.: On the simulation of trailing edge noise with a hybrid LES/APE method. J. Sound and Vibration **270** (2004) 509–524
2. Wagner, S., Bareiß, R., Guidati, G.: Wind Turbine Noise. Springer, Berlin (1996)
3. Howe, M.S.: Trailing edge noise at low mach numbers. J. Sound and Vibration **225** (2000) 211–238

4. Davidson, L., Cokljat, D., Fröhlich, J., Leschziner, M., Mellen, C., Rodi, W.: LESFOIL: Large Eddy Simulation of Flow Around a High Lift Airfoil. Springer, Berlin (2003)
5. El-Askary, W.A.: Zonal Large Eddy Simulations of Compressible Wall-Bounded Flows. PhD thesis, Aerodyn. Inst. RWTH Aachen (2004)
6. Poinsot, T.J., Lele, S.K.: Boundary conditions for direct simulations of compressible viscous flows. J. Comp. Phys. **101** (1992) 104–129
7. Ewert, R., Meinke, M., Schröder, W.: Computation of trailing edge noise via LES and acoustic perturbation equations. Paper 2002-2467, AIAA (2002)
8. Schröder, W., Meinke, M., El-Askary, W.A.: LES of turbulent boundary layers. In: Second International Conference on Computational Fluid Dynamics ICCFD II, Sydney. (2002)
9. El-Askary, W.A., Schröder, W., Meinke, M.: LES of compressible wall bounded flows. Paper 2003-3554, AIAA (2003)
10. Schröder, W., Ewert, R.: Computational aeroacoustics using the hybrid approach (2004) VKI Lecture Series 2004-05: Advances in Aeroacoustics and Applications.
11. Würz, W., Guidati, S., Herr, S.: Aerodynamische Messungen im Laminarwindkanal im Rahmen des DFG-Forschungsprojektes SWING+ Testfall 1 und Testfall 2 (2002) Inst. für Aerodynamik und Gasdynamik, Universität Stuttgart.
12. Strahle, W.C.: Some results in combustion generated noise. J. Sound and Vibration **23** (1972) 113–125
13. Crighton, D., Dowling, A., Williams, J.F.: Modern Methods in analytical acoustics, Lecture Notes. Springer, Berlin (1996)
14. Ewert, R., Schröder, W.: Acoustic perturbation equations based on flow decomposition via source filtering. J. Comp. Phys. **188** (2003) 365–398
15. Bui, T.P., Meinke, M., Schröder, W.: A hybrid approach to analyze the acoustic field based on aerothermodynamics effects. In: Proceedings of the joint congress CFA/DAGA '04, Strasbourg. (2004)
16. Kotake, S.: On combustion noise related to chemical reactions. J. Sound and Vibration **42** (1975) 399–410
17. Germano, M., Piomelli, U., Moin, P., Cabot, W.H.: A dynamic subgrid-scale viscosity model. Phys. of Fluids **7** (1991) 1760–1765
18. Waterson, N.P.: Development of a bounded higher-order convection scheme for general industrial applications. In: Project Report 1994-33, von Karman Institute. (1994)
19. Klein, M., Sadiki, A., Janicka, J.: A digital filter based generation of inflow data for spatially developing direct numerical or large eddy simulations. J. Comp. Phys. **186** (2003) 652–665
20. Düsing, M., Kempf, A., Flemming, F., Sadiki, A., Janicka, J.: Combustion les for premixed and diffusion flames. In: VDI-Berichte Nr. 1750, 21. Deutscher Flammentag, Cottbus. (2003) 745–750
21. Tam, C.K.W., Webb, J.C.: Dispersion-relation-preserving finite difference schemes for computational acoustics. J. Comp. Phys. **107** (1993) 262–281
22. Hu, F.Q., Hussaini, M.Y., Manthey, J.L.: Low-dissipation and low-dispersion runge-kutta schemes for computational acoustics. J. Comp. Phys. **124** (1996) 177–191

Simulation of Vortex Instabilities in Turbomachinery

Albert Ruprecht

Institute of Fluid Mechanics and Hydraulic Machinery, University of Stuttgart, Pfaffenwaldring 10, D-70550 Stuttgart, Germany,
ruprecht@ihs.uni-stuttgart.de

Abstract The simulation of vortex instabilities require a sophisticated modelling of turbulence. In this paper a new turbulence model for Very Large Eddy Simulation is presented. Its main characteristic is an adaptive filtering technique which can distinguish between numerically resolved and unresolved parts of the flow. This unresolved part is then modelled with extended $k-\varepsilon$ model of Chen and Kim. VLES is applied to the simulation of vortex instabilities in water turbines. As a first example the unsteady vortex flows in draft tube is shown and in a second application the unstable flow in a pipe trifurcation is calculated. These cases cannot be predicted accurately with classical turbulence models. Using the new technique, these complex phenomena are well predicted.

Nomenclature

f	$[-]$	filter function
$h_{\max}$	$[\mathrm{m}]$	local grid size
k	$[\mathrm{m^2/s^2}]$	turbulent kinetic energy
L	$[\mathrm{m}]$	Kolmogorov length scale
P_k	$[-]$	production term
u	$[\mathrm{m/s}]$	local velocity
U_i	$[\mathrm{m/s}]$	filtered velocity
$\bar{U_i}$	$[\mathrm{m/s}]$	averaged velocity
$\bar{P}$	$[\mathrm{Pa}]$	averaged pressure
τ_{ij}	$[\mathrm{Pa}]$	Reynolds stresses
α	$[-]$	model constant
Δ	$[\mathrm{m}]$	resolved length scale
Δt	$[\mathrm{s}]$	time step
ε	$[\mathrm{m^2/s^3}]$	dissipation rate
ν	$[\mathrm{m^2/s}]$	kinematic viscosity
ν_t	$[\mathrm{m^2/s}]$	turbulent viscosity
ΔV	$[\mathrm{m^2\ or\ m^3}]$	size of the local element

Subscripts and Superscripts

$\hat{}$	modeled
i	covariant indices, $i = 1, 2, 3$

1 Introduction

The flow in hydraulic turbo machinery is quite complicated, especially under off-design conditions the flow tends too get unsteady and complicated vortex structures occur, which can get unstable. The prediction of these vortex instabilities is quite challenging, since an inaccurate prediction can completely suppress the unsteady motion and result in a steady state flow situation.

It is well known that still one of the fundamental problems of Computational Fluid Dynamics (CFD) is prediction of turbulence. Reynolds-averaged Navier-Stokes (RANS) equations are established as a standard tool for industrial simulations and analysis of fluid flows, although it means that the complete turbulence behaviour has to be enclosed within appropriate turbulence model which takes into account all turbulence scales (from the largest eddies to the Kolmogorov scale). Consequently defining a suitable model for prediction of complex, especially unsteady, phenomena is very difficult.

The highest accuracy for resolving all turbulence scales offers a Direct Numerical Simulation (DNS). It requires a very fine grid and carrying out 3D simulations for complex geometries and flow with high Reynolds number is nowadays time consuming even for high performance computers, Fig. 1. Therefore, DNS is unlikely to be applied to the flow of practical relevance in the near future.

Large Eddy Simulation (LES) starts to be a mature technique for analyzing complex flow, although its major limitation is still expensive computational cost. In the "real" LES all anisotropic turbulent structures are resolved in the computation and only the smallest isotropic scales are modelled. It is schematically shown in Fig. 2. The models used for LES are simple compared to those used for RANS because they only have to describe the influence of the isotropic scales on the resolved anisotropic scales. With increasing Reynolds number the small anisotropic scales strongly decrease becoming isotropic and therefore not

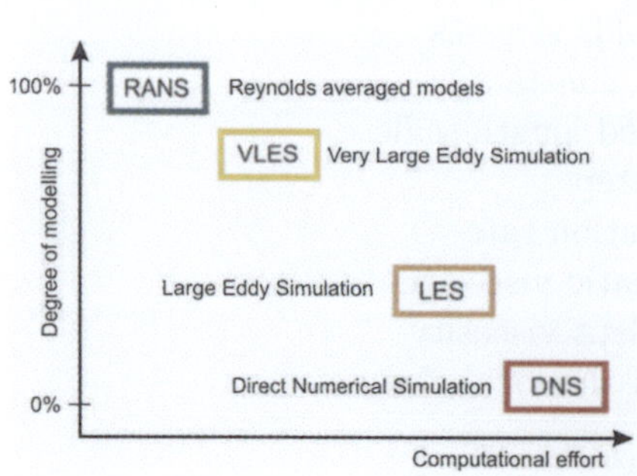

Fig. 1. Degree of turbulence modelling and computational effort for the different approaches

resolvable. There are many "LES" of engineering relevant flows in the literature, although they can be characterised as unsteady RANS (URANS) due to the fact that they only resolve unsteady mean flow not taking into account any turbulence structure.

If there is a gap in the turbulence spectrum between the unsteady mean flow and the turbulent flow, a "classical" RANS i.e. URANS models can be applied, as they are developed for modelling the whole range of turbulent scales, Fig. 2. It also means that they are not suitable for prediction and analysis of many unsteady vortex phenomena.

Contrary, if there is no spectral gap and even one part of the turbulence can be numerically resolved, we can use Very Large Eddy Simulation (VLES). It is very similar to the LES, only that a smaller part of the turbulence spectrum is resolved and the influence of a larger part of the spectrum has to be expressed with the model, see Fig. 2. Nowadays it seems to be a promising compromise for simulation of industrial flow problems with reasonable computational time and costs.

In this paper the development of a VLES turbulence model is presented. It is based on the extended $k-\varepsilon$ model of Chen and Kim [1]. Applying an appropriate filtering technique the new turbulence model distinguishes between resolved and modelled part of the turbulence spectrum. Because of its adaptive characteristic it can be applied for the whole range of turbulence modelling approaches from RANS to DNS.

Here presented applications of the new adaptive turbulence model are simulation of the flow in draft tube and pipe trifurcation. In both cases an unsteady motions are observed and computationally well predicted.

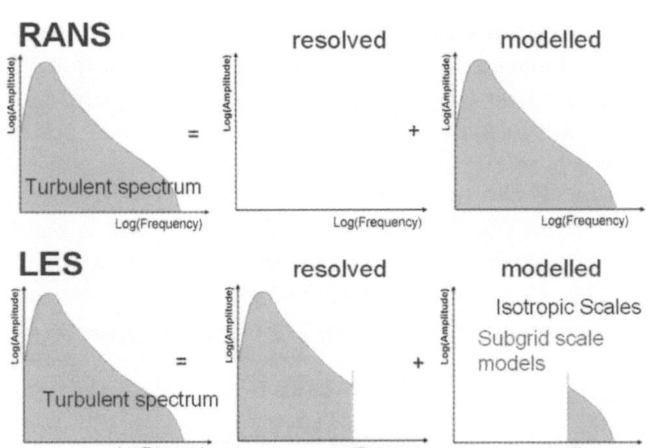

Fig. 2. Modelling approaches for RANS and LES

2 Simulation Method

2.1 Governing Equations

In this work an incompressible fluid with constant properties is considered. The governing equations describing this incompressible, viscous and time dependent flow are the Navier-Stokes equations. They express the conservation of mass and momentum. In the RANS approach, the same equations are time or ensemble averaged leading to the well known RANS equations:

$$\frac{\partial \bar{U}_i}{\partial t} + \bar{U}_j \frac{\partial \bar{U}_i}{\partial x_j} = -\frac{\partial \bar{P}}{\partial x_i} + \nu \nabla^2 \bar{U}_i - \frac{\partial \tau_{ij}}{\partial x_j} \tag{1}$$

$$\frac{\partial \bar{U}_i}{\partial x_i} = 0 \tag{2}$$

In RANS τ_{ij} expresses the Reynolds stress tensor which is unknown and has to be modelled. The task of turbulence modelling is in the formulation and determination of suitable relations for Reynolds stresses. Details of the new VLES approach are described in Sect. 3.

2.2 Numerical Method

The calculations are performed using the program FENFLOSS (Finite Element based Numerical FLOw Simulation System) which is developed at the Institute of Fluid Mechanics and Hydraulic Machinery, University of Stuttgart.

It is based on the Finite Element Method. For spatial domain discretisation 8-node hexahedral elements are used. Time discretisation involves a three-level fully implicit finite difference approximation of 2nd order. For the velocity components and the turbulence quantities a trilinear approximation is applied. The pressure is assumed to be constant within element. For advection dominated flow a Petrov-Galerkin formulation of 2nd order with skewed upwind orientated weighting functions is used.

For the solution of the momentum and continuity equations a segregated algorithm is used. It means that each momentum equation is handled independently. They are linearised and the linear equation system is solved with the conjugated gradient method BICGSTAB2 of van der Vorst [12] with an incomplete LU decomposition (ILU) for preconditioning. The pressure is treated with the modified Uzawa pressure correction scheme [14]. The pressure correction is performed in a local iteration loop without reassembling the system matrices until the continuity error is reduced to a given order.

After solving the momentum and continuity equations, the turbulence quantities are calculated and a new turbulence viscosity is gained. The k and ε-equations are also linearised and solved with BICGSTAB2 algorithm with ILU preconditioning. The whole procedure is carried out in a global iteration until convergence is obtained. For unsteady simulation the global iteration has to be performed for each time step. FENFLOSS flow chart is shown in Fig. 3.

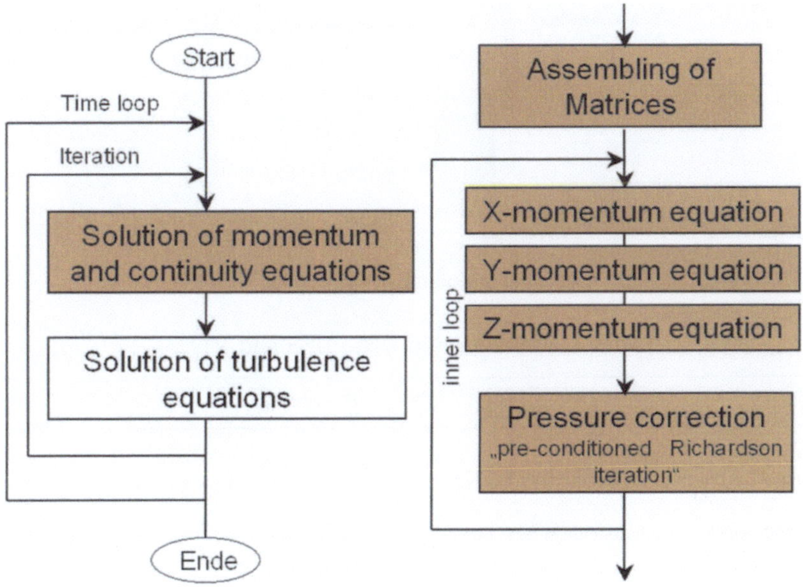

Fig. 3. FENFLOSS flow chart

The code is parallelised and computational domain is decomposed using double overlapping grids. In that case the linear solver BICGSTAB2 has a parallel performance and the data exchange between the domains is organised on the level of the matrix-vector multiplication. The preconditioning is then local on each domain. The data exchange uses MPI (Message Passing Interface) on the computers with distributed memory. On the shared memory computers the code applies OpenMP. For more details on the numerical procedure and parallelisation the reader is referred to [5, 6].

3 Modelling Approach

3.1 Very Large Eddy Simulation

Lately several hybrid methods are proposed in the literature:

- Very Large Eddy Simulation (VLES)
- Semi-Deterministic Simulations (SDS)
- Coherent Structure Capturing (CSC)
- Detached Eddy Simulation (DES)
- Hybrid RANS/LES
- Limited Numerical Scales (LNS)

All of them are based on the same idea to represent a link between RANS and LES. Generally they all can be classified as Very Large Eddy Simulations and

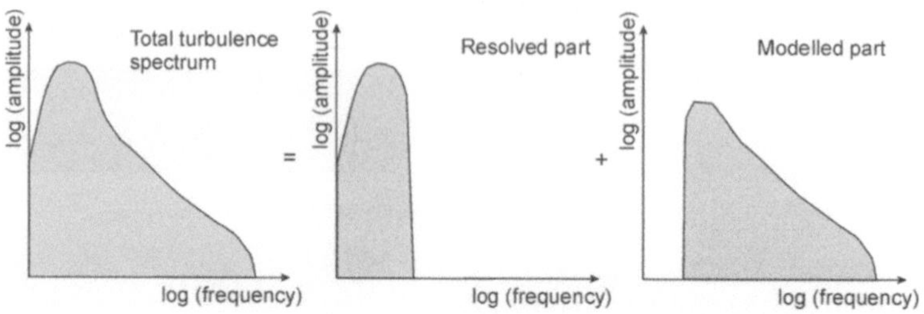

Fig. 4. Modelling approach in VLES

Table 1. Resolution in DNS, LES and VLES [7]

Model	Resolution
Direct numerical simulation (DNS)	All turbulent scales are resolved
Large eddy simulation with near-wall resolution	Grid size and filtering are sufficient to resolve 80% of the energy
Large eddy simulation with near-wall modelling	Grid size and filtering are sufficient to resolve 80% of the energy distant from the wall, but not in the near-wall region
Very large eddy simulation (VLES)	Grid size and filtering are not sufficiently fine to resolve 80% of the energy

their main aim is to overcome the computational costs and capacity problems. These methods try to keep computational efficiency of RANS and the potential of LES to resolve large turbulent structures. Although they can be performed on coarser grids, the simulations are strongly dependent on the modelling.

Above mentioned methods slightly differ in filtering techniques, applied model and interpretation of the resolved motion, but broadly speaking they all have a tendency to solve complex unsteady turbulent flows at high Reynolds number implying a principle "solve less – model more", see Fig. 4 and Table 1. It means that the relevant part of the flow (unsteadiness) is resolved and the rest is modelled.

3.2 Adaptive Turbulence Model

Classical turbulence models, which are usually applied for solving engineering flow problems, model the whole turbulent spectrum. They show excessive viscous behaviour and very often damp down unsteady motion quite early. Therefore they are not completely successful for some flow cases.

VLES is used for resolving at least one part of turbulence spectrum and thus getting more precise picture of the flow behaviour. Depending on the type of the flow and grid size applied the model should automatically adjust to one of the modelling approaches schematically shown in Fig. 5. Therefore an adaptive model is developed. Its advantage is that with increasing computer power it can be afforded that a larger part of spectrum is resolved (due to a finer computational grid). As a result the accuracy of the calculation improves.

For distinguishing resolved and modelled turbulence spectrum (see Fig. 6), the adaptive model uses a filtering technique. There are several of them described in the literature [2, 4, 11], but the applied technique is similar to Willems [13]. The smallest resolved length scale Δ used in filter is according to Magnato and Gabi [4] dependent on the local grid size or the computational time step and local velocity.

The basis of the adaptive model is the $k-\varepsilon$ model of Chen and Kim [1]. It is chosen due to its simplicity and capacity to better handle unsteady flows. Its transport equations for k and ε are given as

$$\frac{\partial k}{\partial t} + \bar{U}_j \frac{\partial k}{\partial x_j} = \frac{\partial}{\partial x_j}\left[\left(\nu + \frac{\nu_t}{\sigma_k}\right)\frac{\partial k}{\partial x_j}\right] + P_k - \varepsilon \tag{3}$$

$$\frac{\partial \varepsilon}{\partial t} + \bar{U}_j \frac{\partial \varepsilon}{\partial x_j} = \frac{\partial}{\partial x_j}\left[\left(\nu + \frac{\nu_t}{\sigma_\varepsilon}\right)\frac{\partial \varepsilon}{\partial x_j}\right] + c_{1\varepsilon}\frac{\varepsilon}{k}P_k - c_{2\varepsilon}\frac{\varepsilon^2}{k} + \underbrace{c_{3\varepsilon}\left[\frac{P_k}{k}\right]\cdot P_k}_{\text{additional term}} \tag{4}$$

with the following coefficients:

$$\sigma_k = 0.75, \quad \sigma_\varepsilon = 1.15, \quad c_{1\varepsilon} = 1.15, \quad c_{2\varepsilon} = 1.9 \quad \text{and} \quad c_{3\varepsilon} = 1.15.$$

Additionally they need to be filtered. According to the Kolmogorov theory it can be assumed that the dissipation rate is equal for all scales. This leads to

$$\varepsilon = \hat{\varepsilon} \tag{5}$$

It is not acceptable for turbulent kinetic energy. Therefore it need filtering

$$\hat{k} = k \cdot \left[1 - f\left(\frac{\Delta}{L}\right)\right]. \tag{6}$$

As a suitable filter

$$f = \begin{cases} 0 & \text{for } \Delta \geq L \\ 1 - \left(\frac{\Delta}{L}\right)^{2/3} & \text{for } L > \Delta \end{cases} \tag{7}$$

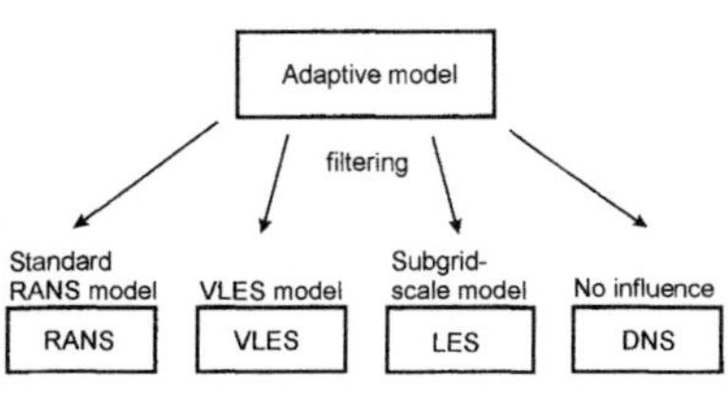

Fig. 5. Adjustment for adaptive model

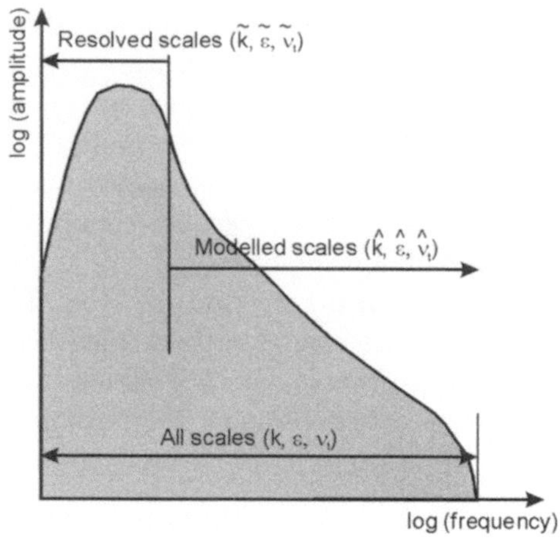

Fig. 6. Distinguishing of turbulence spectrum by VLES

is applied where

$$\Delta = \alpha \cdot \max \begin{cases} |u| \cdot \Delta t \\ h_{\max} \end{cases} \quad \text{with} \quad h_{\max} = \begin{cases} \sqrt{\Delta V} \text{ for 2D} \\ \sqrt[3]{\Delta V} \text{ for 3D} \end{cases} \tag{8}$$

contains model constant α in a range from 1 to 5. Then the Kolmogorov scale L for the whole spectrum is given as

$$L = \frac{k^{3/2}}{\varepsilon}. \tag{9}$$

Modelled length scales and turbulent viscosity are

$$\hat{L} = \frac{\hat{k}^{3/2}}{\hat{\varepsilon}} \tag{10}$$

$$\hat{\nu}_t = c_\mu \cdot \frac{\hat{k}^2}{\hat{\varepsilon}}. \tag{11}$$

with $c_\mu = 0.09$.

The filtering procedure leads to the final equations

$$\frac{\partial k}{\partial t} + U_j \frac{\partial k}{\partial x_j} = \frac{\partial}{\partial x_j}\left[\left(\nu + \frac{\hat{\nu}_t}{\sigma_k}\right)\frac{\partial k}{\partial x_j}\right] + \hat{P}_k - \varepsilon \tag{12}$$

$$\frac{\partial \varepsilon}{\partial t} + U_j \frac{\partial \varepsilon}{\partial x_j} = \frac{\partial}{\partial x_j}\left[\left(\nu + \frac{\hat{\nu}_t}{\sigma_\varepsilon}\right)\frac{\partial \varepsilon}{\partial x_j}\right] + c_{1\varepsilon}\frac{\varepsilon}{k}\hat{P}_k - c_{2\varepsilon}\frac{\varepsilon^2}{k} + c_{3\varepsilon}\left[\frac{\hat{P}_k}{k}\right] \cdot \hat{P}_k \tag{13}$$

with the production term

$$P_k = \nu_t \left(\frac{\partial U_i}{\partial x_j} + \frac{\partial U_j}{\partial x_i}\right)\frac{\partial U_i}{\partial x_j}. \tag{14}$$

For more details of the model and its characteristics the reader is referred to [9].

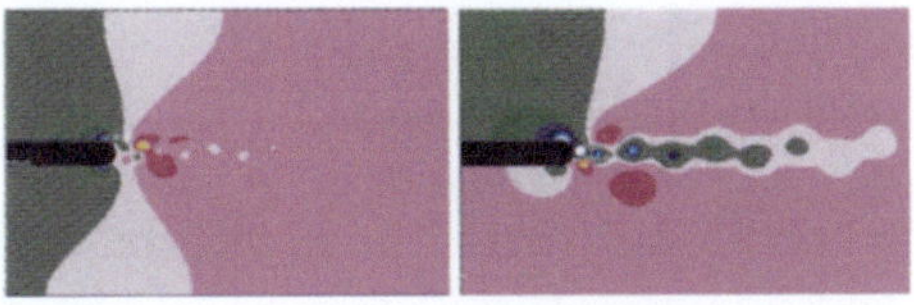

Fig. 7. Pressure distribution by vortex shedding behind the trailing edge, comparison of extended $k-\varepsilon$ model of Chen and Kim and adaptive VLES

The vortex shedding behind the trailing edge, which can be considered as a convenient test case, shows very often difficulties when unsteady RANS is applied. Unsteady RANS with the standard $k-\varepsilon$ model usually leads to a steady state solution. The vortex shedding and its unsteadiness are suppressed by the too diffusive turbulence model. More sophisticated turbulence model i.e. extended $k-\varepsilon$ of Chen and Kim is less diffusive and therefore vortex shedding is gained. Further simulation with VLES method provides slightly improved results. In comparison to $k-\varepsilon$ model of Chen and Kim, it proves to be less damping in the downstream flow behind trailing edge. The comparison of these two models is shown in Fig. 7.

4 Applications

4.1 Swirling Flow in Diffuser and Draft Tube

VLES was used for simulation of swirling flow in a straight diffuser and an elbow draft tube with two piers. In the both cases the specific inlet velocity profile corresponds to the flow at a runner outlet under part load conditions. It is well known that under these conditions an unsteady vortex rope is formed.

Computational grid for straight diffuser had 250000 elements. Applied inlet boundary conditions can be found in [10]. For elbow draft tube two grids were used (180000 and 1 million elements). Computational grid and the inlet boundary conditions (part load operational point of 93%) for the draft tube are shown in Fig. 8.

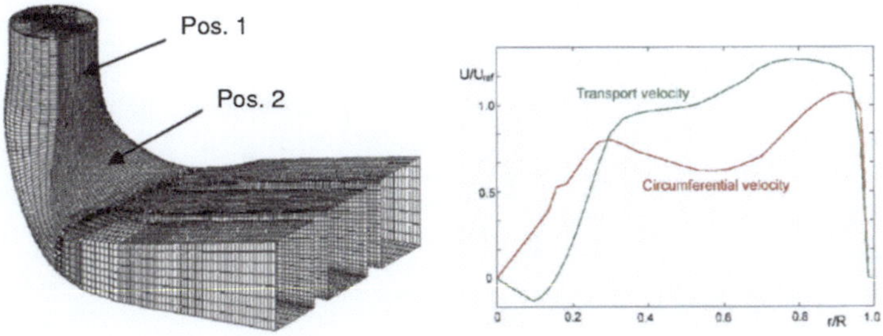

Fig. 8. Computational grid and inlet boundary conditions for the elbow draft tube

Applying an unsteady RANS with the standard $k-\varepsilon$ model leads to a steady state solution. It forms a recirculation region in the center and keeps it steady. On the other hand applying the extended $k-\varepsilon$ of Chen and Kim small unsteady vortex forms. It is too short due to the damping character of the turbulence model. With VLES and adaptive turbulence model the damping of the swirl rate is clearly reduced and vortex rope expends downstream. Comparison of the Chen and Kim model and VLES on example of straight diffuser is shown in Fig. 9.

In a practice elbow draft tubes are usually installed. VLES simulation shows clearly formation of cork-screw type vortex. In Fig. 10 the flow for one time step is shown as well as velocity distribution in a cross section after the bend.

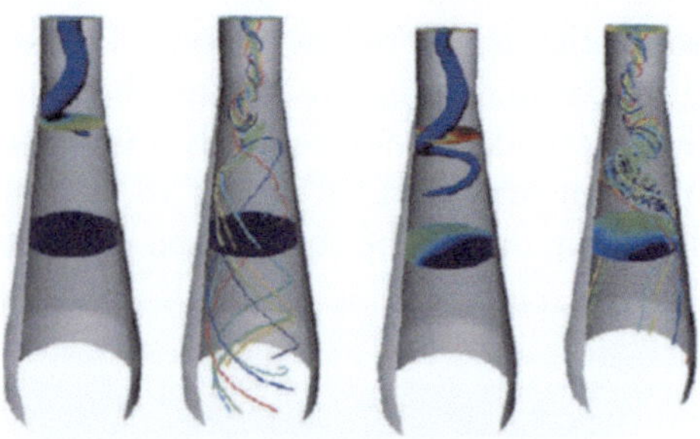

Fig. 9. Vortex rope in a straight diffuser, Chen & Kim (left), VLES (right)

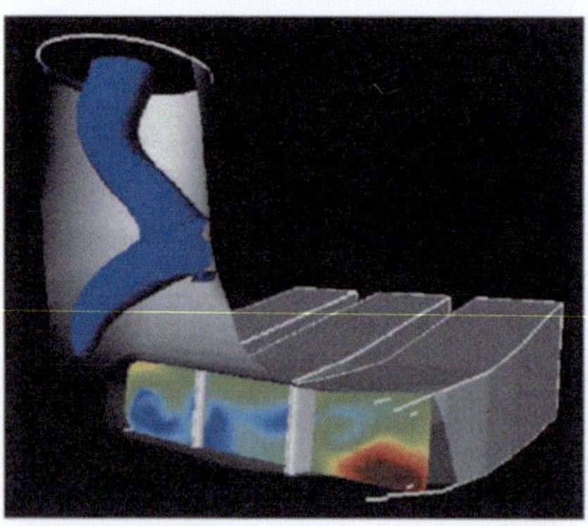

Fig. 10. Simulation of the vortex rope in elbow draft tube with VLES

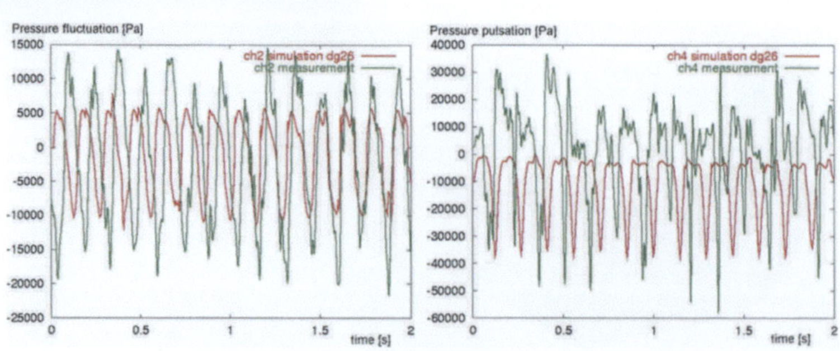

Fig. 11. Comparison of the pressure distribution

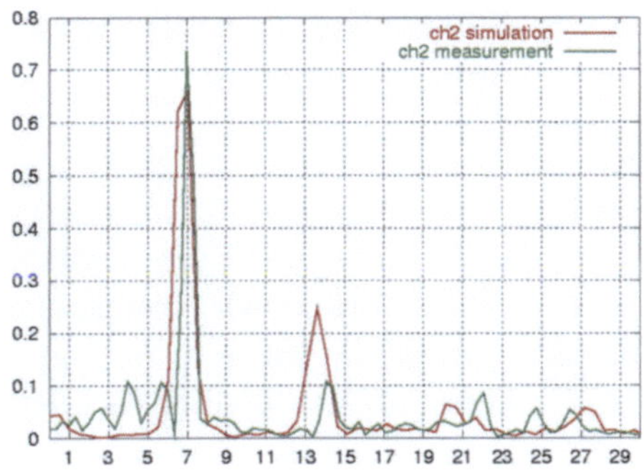

Fig. 12. Fourier transformation of the signals at position 1

Disturbed velocity field can be observed. Therefore, the discharge through the single draft tube channels differs significantly.

The simulation results are compared with experiment data. The pressure distributions at two pints (see Fig. 8) are compared and shown in Fig. 11. It can be seen that the fluctuation amplitudes are higher in the experiment although the frequency corresponds quite well. Fourier transformation of the calculated and measured signal at position 1 is shown in Fig. 12.

4.2 Flow in Pipe Trifurcation

In this section the flow in a pipe trifurcation of a water power plant is presented. The complete water passage consists of the upper reservoir, channel, surge tank, penstock, trifurcation and three turbines. The spherical trifurcation distributes the water from the penstock into the three pipe branches leading to the turbines, Fig. 13. During the power plant exploitation severe power oscillations were encountered at the outer turbines (1 and 3). Vortex instability was discovered as a cause of these fluctuations. The vortex is formed in the trifurcation sphere, appearing at the top and extending into one of the outer branches. After a certain period it changes its behaviour and extends into opposite outer branch. Then the vortex jumps back again. This unstable vortex motion is not periodic and due to its strong swirling flow produces very high losses. These losses reduce the head of the turbine and consequently the power output. For better understand-

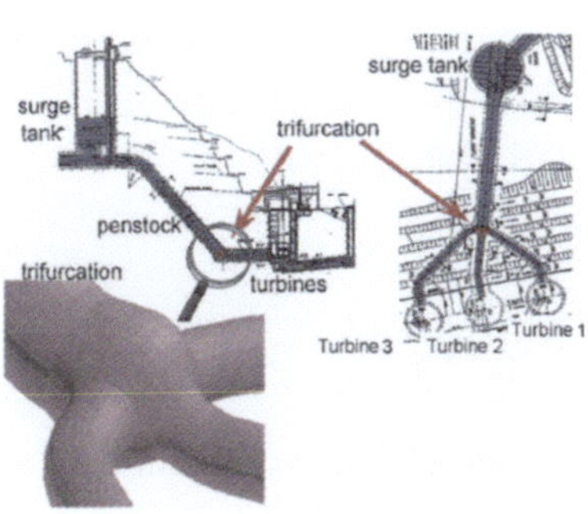

Fig. 13. Water passage with trifurcation

ing and analysis of this flow phenomenon, a computer simulation was performed and results were compared with available model test measurements.

Computational grid had approximately 500000 elements, Fig. 14. Simulation was performed in parallel on 32 processors.

Simulation applying unsteady RANS with the standard $k-\varepsilon$ turbulence model leads to a steady state solution. The obtained vortex structure extends through both outer branches and is fully stable. Thus vortex swirl component is severely underpredicted leading to a poor forecast of the losses in the outer branches. It clearly shows that unsteady RANS is not able to predict this flow phenomenon.

Applying VLES with the new adaptive turbulence model, this unstable vortex movement is predicted. In Fig. 15 the flow inside branch 1 at a certain time step is shown. The vortex is represented by an iso-pressure surface and instantaneous streamlines. After some time (see Fig. 16), vortex "jumps" to the opposite branch. Since the geometry is not completely symmetric, the vortex stays longer in branch 3 than in branch 1. It is observed in the simulation as well as in the model test.

Due to the strong swirl at the inlet of the branch in which the vortex is located, the losses inside this branch are much higher compared to the other two. Therefore, the discharge through this branch is reduced. It is obvious that the discharges i.e. losses through two outer branches vary successively, while the discharge i.e. losses in the middle branch shows much smaller oscillations. In the reality turbines are located at the outlet of each branch. Therefore the discharge variation is rather small since the flow rate through the different branches is

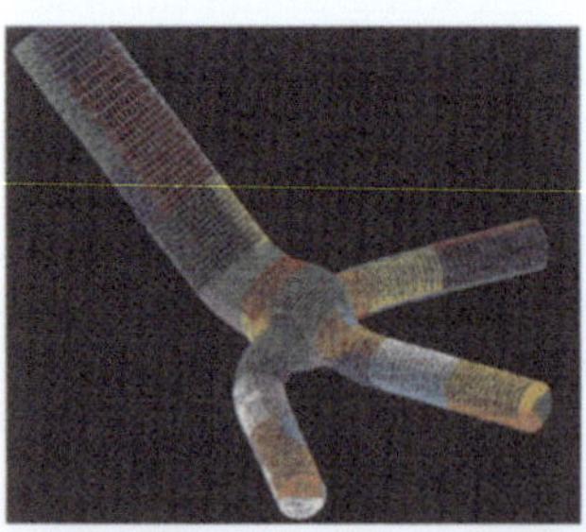

Fig. 14. Computational grid

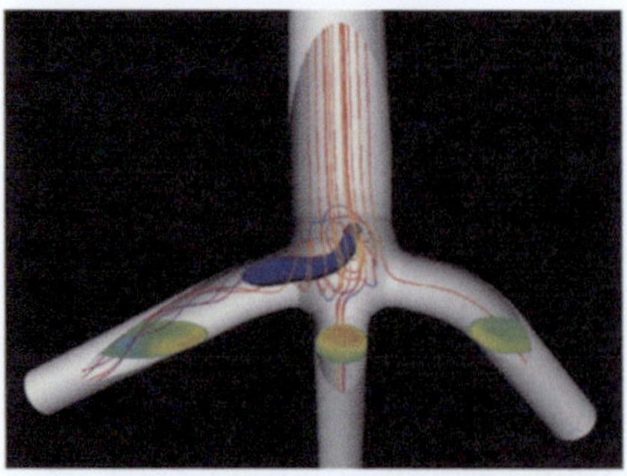

Fig. 15. Flow inside the trifurcation – vortex position in the branch 1

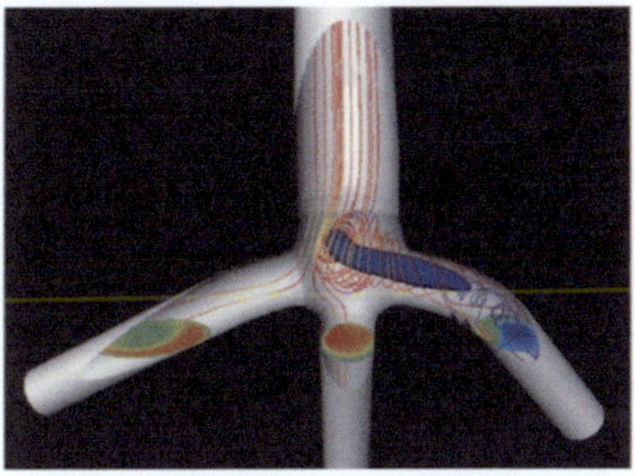

Fig. 16. Flow inside the trifurcation – vortex position in the branch 3

prescribed by the turbines. In the simulation, however, a free outflow boundary condition is applied which leads to the higher discharge variations. For comparison with the experiment, loss coefficients for each branch are calculated, Fig. 17.

Model tests were carried out by ASTRÖ in Graz, Austria. For more details the reader is referred to [3]. The loss coefficients for each branch were calculated from the pressure and discharge measurements. They are shown in Fig. 18.

Comparing the measured loss coefficients with those gained by simulation, it can be seen that the maximum values are still underpredicted, although general flow tendency and quantitative prediction fit in with measurement data reasonably well. This underprediction of the loss coefficient is assumed to be primarily due to the rather coarse grid and secondly due to a strong anisotropic turbulent behaviour which cannot be accurately predicted by the turbulence model based on the eddy viscosity assumption.

In order to solve the oscillation problem in the hydro power plant, it was proposed to change the shape of the trifurcation. To avoid the formation of the

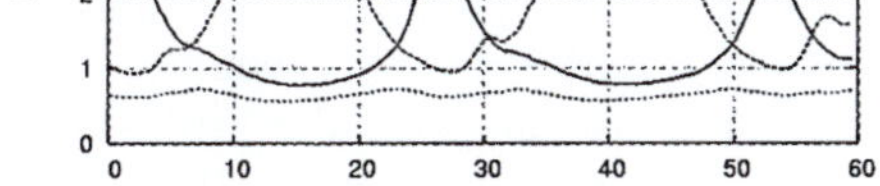

Fig. 17. Loss coefficients for the three branches

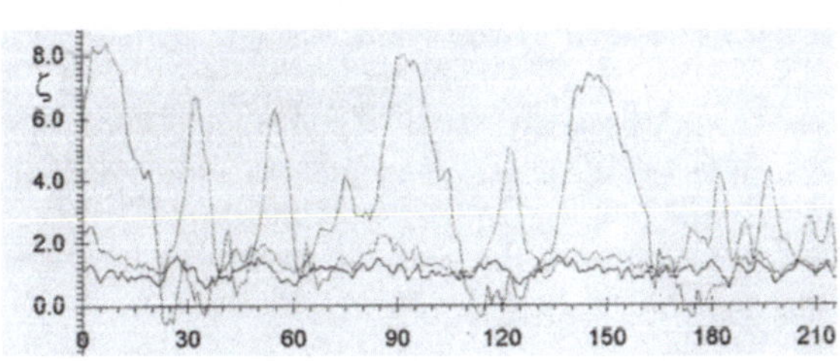

Fig. 18. Measured loss coefficients for all three branches

vortex, upper and lower parts of the sphere are cut off in flat plates. In the meantime this modification was made, the power oscillations disappeared and no unsteady vortex was noticed.

5 Conclusions

An adaptive turbulence model for Very Large Eddy Simulation is presented. It is based on the extended $k-\varepsilon$ model of Chen and Kim. Introducing a filtering technique the new turbulence model distinguishes between numerically resolved and unresolved part of the flow.

With the help of this new model the vortex motions in a draft tube and a pipe trifurcation are calculated. Using classical RANS method and common turbulence models these flow phenomena cannot be predicted. Applying VLES with adaptive turbulence model unsteady vortex motions were obtained due to its less damping character. In all simulated cases the results agree reasonably well with measurement data.

Acknowledgements

The author wants to thank his colleagues Ivana Buntic Ogor, Thomas Helmrich, Ralf Neubauer, who carried out most of the computations.

References

1. Chen, Y.S., Kim, S.W.: Computation of turbulent flows using an extended $k-\varepsilon$ turbulence closure model. NASA CR-179204 (1987)
2. Constantinescu, G.S., Squires, K.D.: LES and DES Investigations of Turbulent flow over a Sphere. AIAA-2000-0540 (2000)
3. Hoffmann, H., Roswora, R.R., Egger, A.: Rectification of Marsyangdi Trifurcation. Hydro Vision 2000 Conference Technical Papers, HCI Publications Inc., Charlotte (2000)
4. Magnato, F., Gabi, M.: A new adaptive turbulence model for unsteady flow fields in rotating machinery. Proceedings of the 8th International Symposium on Transport Phenomena and Dynamics of Rotating Machinery (ISROMAC 8) (2000)
5. Maihöfer, M.: Effiziente Verfahren zur Berechnung dreidimensionaler Strömungen mit nichtpassenden Gittern. Ph.D. thesis, University of Stuttgart (2002)
6. Maihöfer, M., Ruprecht, A.: A Local Grid Refinement Algorithm on Modern High-Performance Computers. Proceedings of Parallel CFD 2003, Elsevier, Amsterdam (2003)
7. Pope, S.B.: Turbulent flows, Cambridge University Press, Cambridge (2000)
8. Ruprecht, A.: Finite Elemente zur Berechnung dreidimensionaler turbulenter Strömungen in komplexen Geometrien, Ph.D. thesis, University of Stuttgart (1989)
9. Ruprecht A.: Numerische Strömungssimulation am Beispiel hydraulischer Strömungsmaschinen, Habilitation thesis, University of Stuttgart (2003)

10. Skotak, A.: Of the helical Vortex in the Turbine Draft Tube Modelling. Proceedings of the 20th IAHR Symposium on Hydraulic Machinery and Systems, Charlotte, USA (2000)
11. Spalart, P.R., Jou, W.H., Strelets, M., Allmaras, S.R.: Comments on the Feasibility of LES for Wings, and on Hybrid RANS/LES Approach. In: Liu C., Liu Z. (eds.) Advances in DNS/LES, Greyden Press, Columbus (1997)
12. van der Vorst, H.A.: Recent Developments in Hybrid CG Methods. In: Gentzsch, W., Harms, U. (eds.) High-Performance Computing and Networking, vol. 2: Networking and Tools, Lecture Notes in Computer Science, Springer, **797** (1994) 174–183
13. Willems, W.: Numerische Simulation turbulenter Scherströmungen mit einem Zwei-Skalen Turbulenzmodell, Ph.D. thesis, Shaker Verlag, Aachen (1997)
14. Zienkiewicz, O.C., Vilotte, J.P., Toyoshima, S., Nakazawa, S.: Iterative method for constrained and mixed finite approximation. An inexpensive improvement of FEM performance. Comput Methods Appl Mech Eng **51** (1985) 3–29

Applications III: Particle Methods

Atomistic Simulations on Scalar and Vector Computers

Franz Gähler[1] and Katharina Benkert[2]

[1] Institute for Theoretical and Applied Physics, University of Stuttgart, Pfaffenwaldring 57, D-70550 Stuttgart, Germany, gaehler@itap.physik.uni-stuttgart.de

[2] High Performance Computing Center Stuttgart (HLRS), Nobelstraße 19, D-70569 Stuttgart, Germany, benkert@hlrs.de

Abstract Large scale atomistic simulations are feasible only with classical effective potentials. Nevertheless, even for classical simulations some ab-initio computations are often necessary, e.g. for the development of potentials or the validation of the results. Ab-initio and classical simulations use rather different algorithms and make different requirements on the computer hardware. We present performance comparisons for the DFT code VASP and our classical molecular dynamics code IMD on different computer architectures, including both clusters of microprocessors and vector computers. VASP performs excellently on vector machines, whereas IMD is better suited for large clusters of microprocessors. We also report on our efforts to make IMD perform well even on vector machines.

1 Introduction

For many questions in materials science, it is essential to understand dynamical processes in the material at the atomistic level. Continuum simulations cannot elucidate the dynamics of atomic jump processes in diffusion, in a propagating dislocation core, or at a crack tip. Even for many static problems, like the study of the structure of grain boundaries, atomistic simulations are indispensable. The tool of choice for such simulations is molecular dynamics (MD). In this method, the equations of motion of a system of interacting particles (atoms) are directly integrated numerically. The advantage of the method is that one needs to model only the interactions between the particles, not the physical processes to be studied. The downside to this is a high computational effort.

The interactions between atoms are governed by quantum mechanics. Therefore an accurate and reliable simulation would actually require a quantum mechanical model of the interactions. While this is possible in principle, in practice it is feasible only for rather small systems. Computing the forces by ab-initio density functional theory (DFT) is limited to a few hundred atoms at most, especially if many transition metal atoms with a complex electronic structure are part of the system. For ab-initio MD, where many time steps are required, the limits are even much smaller. Due to the bad scaling with the number of atoms (N^3 for part of the algorithm), there is little hope that one can exceed these limits in the foreseeable future. Order N algorithms, which are being studied for insulators, do not seem to be applicable to metal systems.

For many simulation problems, however, systems with a few hundred atoms are by far not big enough. Especially the study of mechanical processes, like dislocation motion, crack propagation, or nano-indentation would at least require multi-million atom systems. Such simulations are possible only with classical effective potentials. These must be carefully fitted to model the quantum mechanical interactions as closely as possible. One way to do this is by force matching [1]. In this method, for a collection of small reference structures, which should comprise all typical local configurations, the forces on all particles are computed quantum-mechanically, along with other quantities like energies and stresses. The effective potentials are then fitted to reproduce these reference forces. This procedure is well known for relatively simple materials, but has successfully been applied recently also to complex intermetallics [2]. Force matching provides a way to bridge the gap between the requirements of large scale MD simulations and what is possible with ab-initio methods, thus making quantum mechanical information available also to large scale simulations.

For accurate and reliable simulations of large systems, both classical and quantum simulations are necessary. The quantum simulations are needed not only for the development of effective potentials, but also for the validation of the results. The two kinds of simulations use rather different algorithms, and have different computer hardware requirements. If geometric domain decomposition is used, classical MD with short range interactions is known to scale well to very large particle and CPU numbers. It also performs very well on commodity microprocessors. For large simulations, big clusters of such machines, together with a low latency interconnect, are therefore the ideal choice. On the other hand, vector machines have the reputation of performing poorly on such codes.

With DFT simulations, the situation is different; they do not scale well to very large CPU numbers. Among other things this is due to 3D fast Fourier transforms (FFT) which takes about a third of the computation time. It is therefore important to have perhaps only a few, but very fast CPUs, rather than many slower ones. Moreover, the algorithms do mostly linear algebra and need, compared to classical MD, a very large memory. Vector machines like the NEC SX series therefore look very promising for the quantum part of the simulations.

The remainder of this article is organized into three parts. In the first part, we will analyze the performance of VASP [3, 4, 5], the Vienna Ab-initio Simulation Package, on the NEC SX and compare it to the performance on a powerful microprocessor based machine. VASP is a widely used DFT code and is very efficient for metals, which we are primarily interested in. In the second part, the algorithms and data layout of our in-house classical MD code IMD [6] are discussed and performance measurements on different cluster architectures are presented. In the third part, we describe our efforts to achieve competitive performance with classical MD also on vector machines. So far, these efforts have seen only a limited success.

2 Ab-initio Simulations with VASP

The Vienna Ab-initio Simulation Package, VASP [3, 4, 5], is our main work horse for all ab-initio simulations. In recent years, its development has been concentrated on PC clusters, where it performs very well, but the algorithms used should also perform well on vector machines. As explained above, due to the modest scaling with increasing CPU numbers it is very important to have fast CPUs available. Vector computing is therefore the obvious option to explore. For these tests an optimized VASP version for the NEC SX has been used.

As test systems, we take two large complex metal systems: $Cd_{186}Ca_{34}$ with 220 atoms per unit cell and $Cd_{608}Ca_{104}$ with 712 atoms per unit cell. In each case, one electronic optimization was performed, which corresponds to one MD step. As we explain later, the runtimes for such large systems are too big to allow for a large number of steps. However, structure optimizations through relaxation simulations are possible. In all cases, k-space was sampled at the Γ-point only. Two VASP versions were used: a full complex version and a specialized Γ-point only version. The latter uses a slightly different algorithm which is faster and uses less memory, but can be used only for the Γ-point. Timings are given in Table 1. For comparison, also the timings on an Opteron cluster are included. These timings show that the vector machine has a clear advantage compared to a fast microprocessor machine. Also the absolute gigaflop rates are very satisfying, reaching up to 55% of the peak performance for the largest system.

The scaling with the number of CPUs is shown in Fig. 1. As can be seen, the full complex version of VASP scales considerably better. This is especially true for the SX8, which shows excellent scaling up to 8 CPUs, whereas for the SX6+ the performance increases subproportionally beyond 6 CPUs. For the Γ-point only version, the scaling degrades beyond 4 CPUs, but this version is still faster than the full version. If only the Γ-point is needed, it is worthwhile to use this version.

Table 1. Timings for three large systems on the SX8 (with SX6 executables), the SX6+, and an Opteron cluster (2GHz, Myrinet). For the vector machines, both the total CPU time (in seconds) and the gigaflop rates are given

	SX8		SX6+		Opteron
	time	GF	time	GF	time
712 atoms, 8 CPUs, complex	47256	70	88517	38	
712 atoms, 8 CPUs, Γ-point	13493	57	20903	36	70169
220 atoms, 4 CPUs, complex	2696	33	5782	15	13190

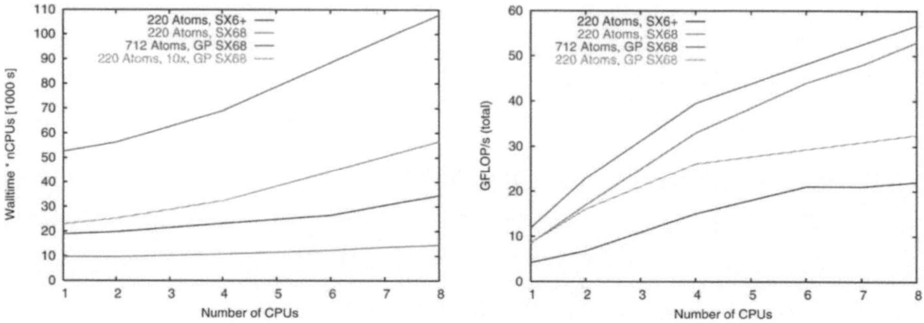

Fig. 1. Scaling of VASP for different systems on the SX8 (with SX6 executables) and the SX6+. Shown are total CPU times (left) and absolute gigaflop rates (right). The timings of the 220 atom system (Γ-point only version) on the SX8 have been multiplied by 10

3 Classical Molecular Dynamics with IMD

For all classical MD simulations we use our in-house code, IMD [6]. It is written in ANSI C, parallelized with MPI, and runs efficiently on a large number of different hardware architectures. IMD supports many different short range interactions for metals and covalent ceramics. Different integrators and a number of other simulation options are available, which allow, e.g., to apply external forces and stresses on the sample. In the following, we describe only those parts of the algorithms and data layout, which are most relevant for the performance. These are all concerned with the force computation, which takes around 95% of the CPU time.

3.1 Algorithms and Data Layout

If the interactions have a finite range, the total computational effort of an MD step scales linearly with the number of atoms in the system. This requires, however, to quickly find those (few) atoms from a very large set, with which a given atom interacts. Searching the whole atom list each time is an order N^2 operation, and is not feasible for large systems. For moderately big systems, Verlet neighbor lists are often used. The idea is to construct for each atom a list of those atoms which are within the interaction radius r_c, plus an extra margin r_s (the skin). The construction of the neighbor lists is still an order N^2 operation, but depending on the value of r_s they can be reused for a larger or smaller number of steps. The neighbor lists remain valid as long as no atom has traveled a distance larger than $r_s/2$.

For very large systems, Verlet neighbor lists are still not good enough and link cells are usually used. In this method, the system is subdivided into cells, whose diameter is just a little bigger than the interaction cutoff. Atoms can then interact only with atoms in the same and in neighboring cells. Sorting the atoms into the cells is an order N operation, and finding the atoms in the same and in neighboring cells is order N, too. In a parallel simulation, the sample is simply divided into blocks of cells, each CPU dealing with one block (Fig. 2). Each block is surrounded by a layer of buffer cells, which are filled before the force computation with copies of atoms on neighboring CPUs, so that the force can be computed completely locally.

This algorithm, which is manifestly of order N, is fairly standard for large scale MD simulations. Its implementation in IMD is somewhat special in one respect. The cells store the whole particle data in per-cell arrays and not indices into a big array of all atoms. This has the advantage that nearby atoms are stored closely together in memory as well, and stay close during the whole simulation. This is a considerable advantage on cache-based machines. The price to pay is an extra level of indirect addressing, which is a disadvantage on vector machines.

Although the link cell algorithm is of order N, there is still room for improvement. It can in fact be combined with Verlet neighbor lists. The advantage of doing this is explained below. The number of atoms in a given cell and its neighbors is roughly proportional to $(3r_c)^3$, where r_c is the interaction cutoff radius. In the link cell algorithm, these are the atoms which are potentially interacting with a given atom, and so at least the distance to these neighbors has to be computed. However, the number of atoms within the cutoff radius is only proportional to $\frac{4\pi}{3}r_c^3$, which is by a factor $\frac{81}{4\pi} \approx 6.45$ smaller. If Verlet lists are used, a large number of these distance computations can be avoided. The link cells are then used only to compute the neighbor lists (with an order N method), and the neighbor lists are used for the force computations. This leads to a runtime reduction of 30–40%, depending on the machine and the interaction model (the simpler the interaction, the more important the avoided distance computations). The downside of using additional neighbor lists is a substantially

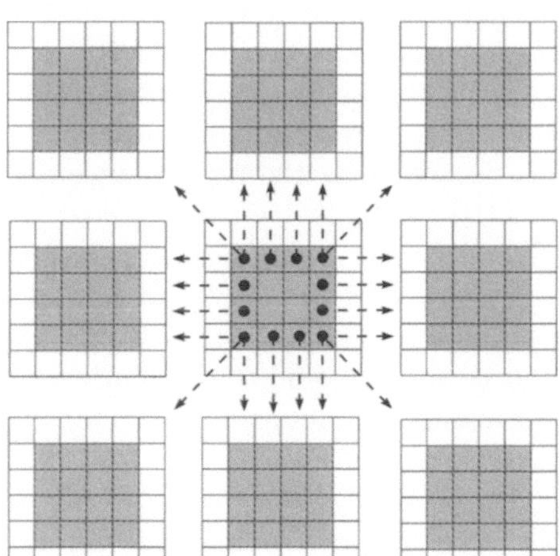

Fig. 2. Decomposition of the sample into blocks and cells. Each CPU deals with one block of cells. The white buffer cells contain copies of cells on neighbor CPUs, so that forces can be computed locally

increased memory footprint. On systems like the Cray T3E, neighbor lists are therefore not feasible, but on today's cluster systems they are a very worthwhile option.

There is one delicate point to be observed, however. If any atom is moved from one cell to another, or from one CPU to another, the neighbor lists are invalidated. As this could happen in every step *somewhere* in the system, these rearrangements of the atom distribution must be postponed until the neighbor tables have to be recomputed. Until then, atoms can leave their cell or CPU at most by a small amount $r_s/2$, which does not matter. The neighbor tables contain at each time all interacting neighbor particles.

3.2 Performance Measurements

We have measured the performance and scaling of IMD on four different cluster systems: a HP XC6000 cluster with 1.5 GHz Itanium processors and Quadrics interconnect, a 3.2 GHz Xeon EM64T cluster with Infiniband interconnect, a 2 GHz Opteron cluster with Myrinet 2000 interconnect, and an IBM Regatta cluster (1.7 GHz Power4+) with IBM High Performance Switch (Figs. 3–4). Shown is the CPU time per step and atom, which should ideally be a horizontal line. On each machine, systems of three sizes and with two different interactions are simulated. The systems have about 2k, 16k, and 128k atoms per CPU. One system is an FCC crystal interacting with Lennard-Jones pair interactions, the other a B2 NiAl crystal interacting with EAM [7] many-body potentials. The different system sizes probe the performance of the interconnect: the smaller the system per CPU, the more important the communication, especially the latency. As little as 2000 atoms per CPU is already very demanding on the interconnect.

The fastest machine is the Itanium system, with excellent scaling for all system sizes. For the smallest systems and very small CPU number, the performance increases still further, which is probably a cache effect. This performance was not easy to achieve, however. It required careful tuning and some rewriting of the innermost loops (which do not harm the performance on the other machines). Without these measures the code was 3–4 times slower, which would not be acceptable. Unfortunately, while the tuning measures had the desired effect with the Intel compiler releases 7.1, 8.0, and 8.1 up to 8.1.021, they do not seem to work with the newest releases 8.1.026 and 8.1.028, with which the code is again slow. So, achieving good performance on the Itanium is a delicate matter.

The next best performance was obtained on the 64bit Xeon system. Its Infiniband interconnect also provides excellent scaling for all system sizes. One should note, however, that on this system we could only use up to 64 processes, because the other nodes had hyperthreading enabled. With hyperthreading it often happens that both processes of a node run on the same physical CPU, resulting in a large performance penalty. For a simulation with four processes per node, there was not enough memory, because the Infiniband MPI library allocates buffer space in each process for every other process.

The Opteron system also shows excellent performance, but only for the two larger system sizes. The small systems seem to suffer from the interconnect latency. The performance penalty saturates, however, at about 20%. We should also mention that these measurements have been made with binaries compiled with gcc. We expect that using the PathScale or Intel compilers would result in a 5–10% improvement.

Finally, the IBM regatta system is the slowest of the four, but also shows excellent scaling for all system sizes. For very small CPU numbers, the performance was a bit erratic, which may be due to interferences with other processes running on the same 32 CPU node.

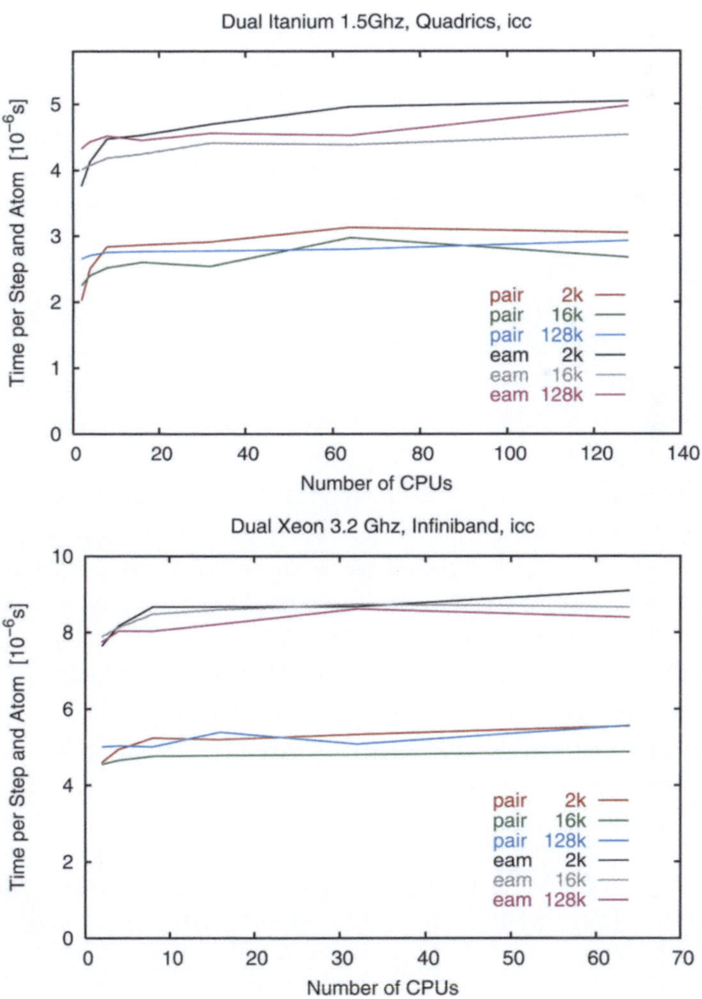

Fig. 3. Scaling of IMD on the Itanium (top) and Xeon (bottom) systems

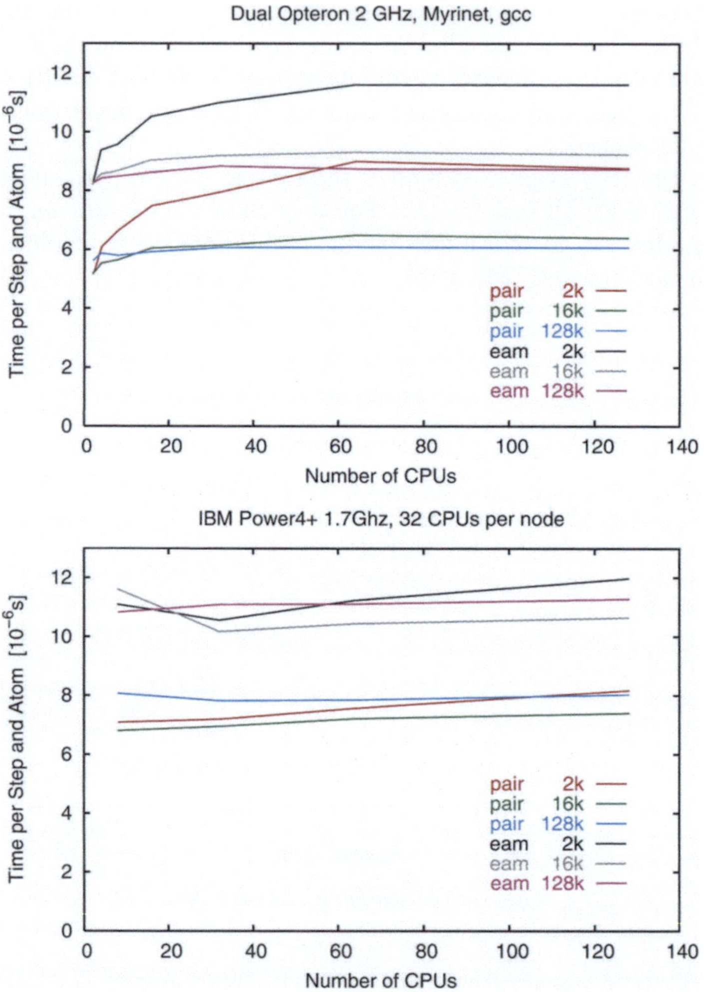

Fig. 4. Scaling of IMD on the Opteron (top) and IBM Regatta (bottom) systems

4 Classical Molecular Dynamics on the NEC SX

The algorithm for the force computation sketched in Sect. 3.1 suffers from two problems, when executed on vector computers. The innermost loop over interacting neighbor particles is usually too short, and the storage of the particle data in per-cell arrays leads to an extra level of indirect addressing. The latter problem could be solved in IMD by using a different memory layout for the vector version, in which the particle data is stored in single big arrays and not in per-cell arrays. The cells then contain only indices into the big particle list. In order to keep as much code as possible in common between the vector and the scalar versions of IMD, all particle data is accessed via preprocessor macros.

The main difference between the two versions of the code is consequently the use of two different sets of access macros. The problem of the short loops has to be solved by a different loop structure. We have experimented with two different algorithms, the Layered Link Cell (LLC) algorithm [8], and the Grid Search algorithm [9].

4.1 The LLC Algorithm

The basic idea of the LLC algorithm [8] is to divide the list of all interacting atom pairs (implicitly contained in the Verlet neighbor list) into blocks of independent atom pairs. The pairs in a block are independent in the sense, that no particle occurs twice at the first position of the pairs in the block, nor twice at the second position. After all the forces between the atom pairs in a block have been computed, they can be added in a first loop to the particles at the first position, and in a second loop to the particles at the second position. Both loops are obviously vectorizable.

The blocks of independent atom pairs are constructed as follows. Let m be the maximal number of atoms in a cell. The set of particles at the first position of the pairs in the block is simply the set of all particles. The particle at position i in cell q is then paired with particle $i + k \bmod m$ in cell q', where q' is a cell at a fixed position relative to q (e.g., the cell just to the right of q), and k is a constant between 0 and m (0 is excluded, if $q = q'$). For each value of the neighbor cell separation and constant k, an independent block of atom pairs is obtained.

Among the atom pairs in the lists constructed above, there are of course many which are too far apart to be interacting. The lists are therefore reduced to those pairs, whose atoms have a distance not greater than $r_c + r_s$. These reduced pair lists replace the Verlet neighbor lists, and remain valid as long as no particle has traveled a distance larger than $r_s/2$, so that they need not be recomputed at every step.

The algorithm just described has been implemented in IMD, but its performance on the NEC SX is still modest (see Sect. 4.3). One limitation of the LLC algorithm is certainly that it requires the cells to have approximately the same number of atoms. Otherwise, the performance will degrade substantially. This condition was satisfied, however, by our crystalline test systems. In order to understand the reason for the modest performance, we have reimplemented the algorithm afresh, in a simple environment instead of a production code, both in Fortran 90 and in C. It turned out that the C version performs similarly to IMD, whereas the Fortran version is about twice as fast on the NEC SX (Sect. 4.3). The Fortran compiler apparently optimizes better than the C compiler.

4.2 The Grid Search Algorithm

As explained in Sect. 3.1, most of the particles in neighboring cells are too far away from a given one in the cell at the center to be interacting. This originates from the fact that a cube poorly approximates a sphere, especially if the cube has edge length 1.5 times the diameter of the sphere, as it is dictated by the link cell algorithm. The resulting, far too many distance computations can be avoided to some extent using Verlet neighbor lists, but only an improved version of the LLC algorithm (the Grid Search algorithm) presents a true solution to this problem.

If one would use smaller cells, the sphere of interacting particles could be approximated much better. However, this would result in a larger number of singly occupied or empty cells, making it very inefficient to find interacting particles. A further problem is, that with each cell a certain bookkeeping overhead is involved. As the number of cells would be much larger, this cost is not negligible, and should be avoided.

The Grid Search algorithm tries to combine the advantages of a coarse and a fine cell grid, and avoids the respective disadvantages.

The initial grid is relatively coarse, having 2–3 times more cells than particles. To use a simplified data structure, we demand at most one particle per cell, a precondition which cannot be guaranteed in reality. In case of multiply occupied cells, particles are reassigned to neighboring cells using *neighbor cell assignment* (NCA). This keeps the number of empty cells to a minimum. During NCA each particle gets a virtual position in addition to its true position. To put it forward in a simple way, the virtual positions of particles in multiply occupied cells are iteratively modified by shifting these particles away from the center of the cell on the ray connecting the center of the cell and the particle's true position. As soon as the precondition is satisfied the virtual positions are discarded. Only the now compliant assignment of particle to cell, stored in a one-dimensional array, and the largest virtual displacement d_{max}, denoting the maximal distance between the virtual and true position of all particles, are kept.

The so-called *sub-cell grouping* (SCG) exploits the exact positions of the particles relative to their cells by introducing a finer hierarchical grid. This reduces the number of unnecessarily examined particle pairs and distance calculations. To simplify the explanation, we assume in first instance that NCA is not used.

The basic idea of Grid Search is to palter with chance to get a "successful" distance computation. We consider a pair of two cells, the cell at the center C and a neighbor cell N, with one particle located in each cell. In the convenient case, the neighbor cell is sufficiently close to the cell at the center (Fig. 5), so that there is a good chance that the two particles contained in the cells are interacting.

In the complicated case, if the neighbor cell is so far apart of the cell at the center (Fig. 6) that there is only a slight chance that the particle pair gets inserted into the Verlet list, SCG comes into play. The cell at the center is divided into a number of sub-cells, depending on integer arithmetic. Extra sub-cells are added for particles that have been moved by NCA to neighboring cells for each quadrant/octant (Fig. 7). A fixed sub-cell/neighbor cell relation is denoted as *group*.

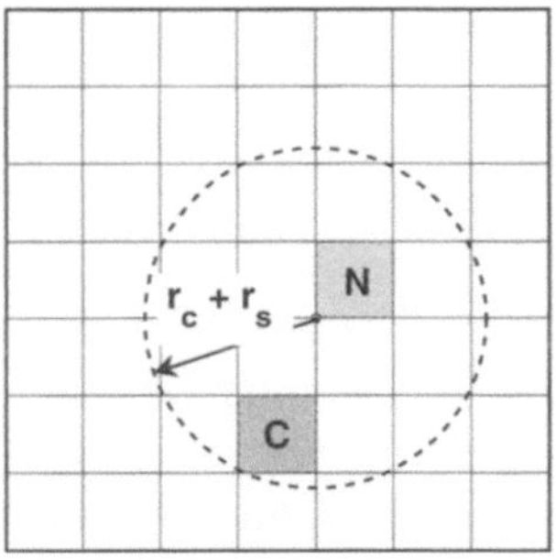

Fig. 5. Cell at the center C is sufficiently close to neighbor cell N

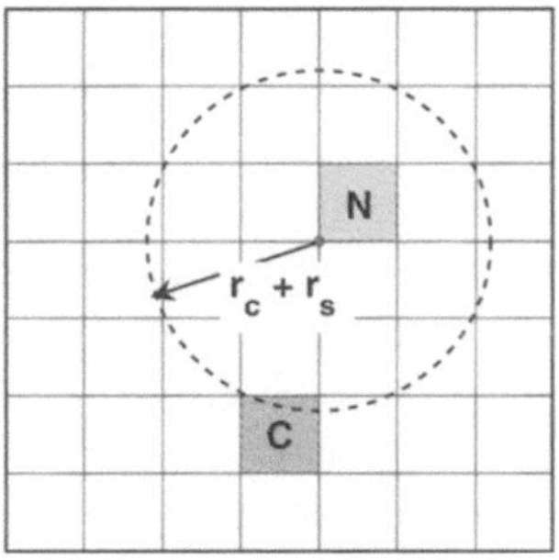

Fig. 6. Cell at the center C is not close enough to neighbor cell N

By comparing the minimal distance between each sub-cell and the neighbor cell to $r_c + r_s$, a number of groups can be excluded in advance. As shown in Fig. 8, only $\frac{1}{4}$ of the initial cell at the center needs to be searched.

The use of NCA complicates SCG, because it changes the condition for excluding certain groups for a given neighbor cell relation in advance: the minimal distance between a sub-group and a neighbor cell does no longer have to be smaller or equal than $r_c + r_s$, but smaller or equal than $r_v = r_c + r_s + d_{max}$. The virtual displacement occurs only once in r_v, since one particle is known to be located in the sub-cell, and the other one can be displaced by as much as d_{max}. Thus, the set of groups that need to be considered changes whenever the particles are redistributed into the cells, i.e., whenever the Verlet list is updated.

In order to reduce the amount of calculations and to save memory, a data structure is established, stating whether a given group can contain interacting particles for a certain virtual displacement. For 32(64)-bit integer arithmetics, the cell at the center is divided into $4 \times 4 \times 3$ ($3 \times 3 \times 2$) sub-cells and eight extra-cells (one for each octant) resulting in 56(26) groups. So in a two-dimensional integer array, the first dimension being the neighbor cell relation, the second indicating a certain pre-calculated value of d_{max}, the iGr-th bit (iGr is the group

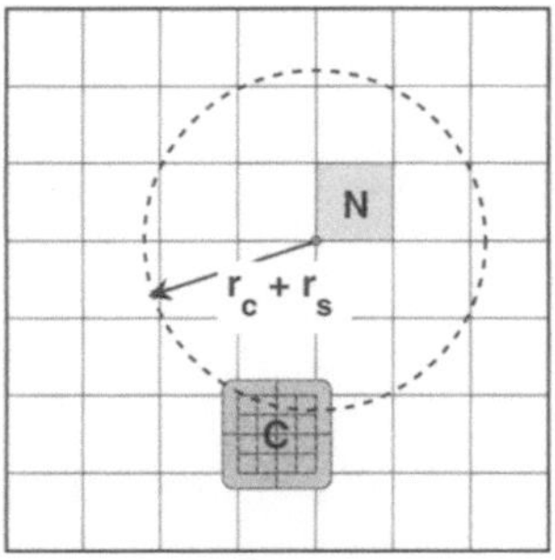

Fig. 7. Cell at the center is divided into sub-cells

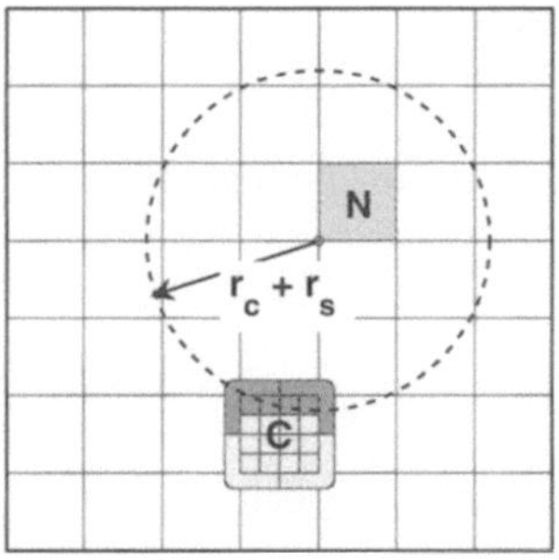

Fig. 8. Some groups can be excluded from search

number) is set to 1 if the minimal distance between the sub-cell and the neighboring cell is not greater than r_v.

The traditional LLC data structures, a one-dimensional array with the number of particles in each cell and a two-dimensional array listing the particles in each cell, are used in Grid Search on the sub-cell level: a one-dimensional array storing the number of particles in each group and a two-dimensional array listing the particles in each group. Together with the array of cell inhabitants produced by the NCA, this represents a double data structure on cell and sub-cell level, respectively: for each cell we know the particle located in it, and for each sub-cell we know the total number and which particles are located in it.

As in the LLC algorithm, independent blocks of the Verlet list consist of all particle pairs having a constant neighbor cell relation. The following code examples describe the setup of the Verlet list. For neighbor cells sufficiently close to the cell at the center, the initial grid is used:

```
do for all particles j1
  if the neighbor cell of the cell with particle j1 contains
    a~particle j2 then
      save particles to temporary lists
  endif
end do
```

If the distance of the neighbor cell to the cell at the center is close to r_v, then SCG is used:

```
do for all sub-cells
  if particles in this sub-cell and the given neighbor cell
  can interact then
    do for all particles in this sub-cell
      if the neighbor cell of the sub-cell with particle j1
        contains a~particle j2 then
          save particles to temporary lists
      endif
    end do
  end if
end do
```

The temporary lists are then, as in the LLC algorithm, reduced to those pairs whose atoms have a distance not greater than $r_c + r_s$.

4.3 Performance Measurements

To compare the performance of the LLC and the Grid Search (GS) algorithms, an FCC crystal with 16384 or 131072 atoms with Lennard-Jones interactions is simulated over 1000 time steps using a velocity Verlet integrator. As reference, the same system has also been simulated with the LLC algorithm as implemented in IMD. The execution times are given in Fig. 9. Not shown is the reimplementation of the LLC algorithm in C, which shows a similar performance as IMD.

For the Grid Search algorithm, the time per step and atom is about 1.0 μs, which is more than twice as fast as IMD on the Itanium system. However, such a comparison is slightly unfair. The Itanium machine simulated a system with two atom types and a tabulated Lennard-Jones potential, which could be replaced by any other potential without performance penalty. The vector version, in contrast, uses computed Lennard-Jones potentials and only one atom type (hard-coded), which is less flexible but faster. Moreover, there was no parallelization overhead. When simulating the same systems as on the Itanium with IMD on the NEC SX8, the best performance obtained with the 128k atom sample resulted in 2.5 μs per step and atom. This is roughly on par with the Itanium machine. An equivalent implementation of Grid Search in Fortran would certainly be faster, but probably by a factor of less than two.

Next, we compare the performance on the NEC SX6+ and the new NEC SX8. The speedup of an SX6+ executable running on SX8 should theoretically be 1.78, since the SX6+ CPU has a peak performance of 9 GFlop/s, whereas the SX8 CPU has 16 GFlop/s. Recompiling on SX8 may lead to even faster execution times, benefiting e.g. from the hardware square root or an improved data access with stride 2.

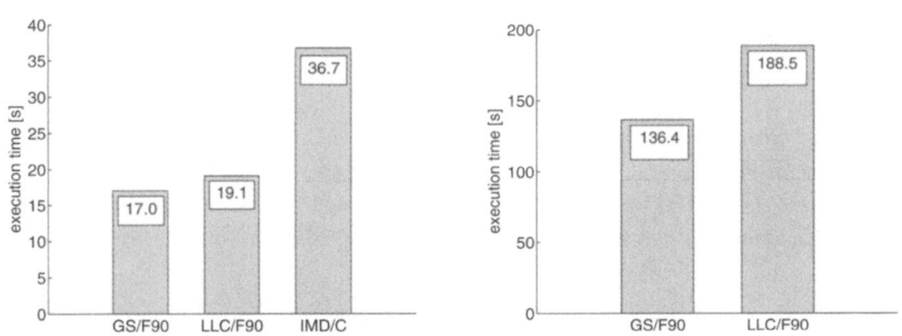

Fig. 9. Execution times of the different algorithms on the NEC SX8, for FCC crystals with 16k atoms (left) and 131k atoms (right)

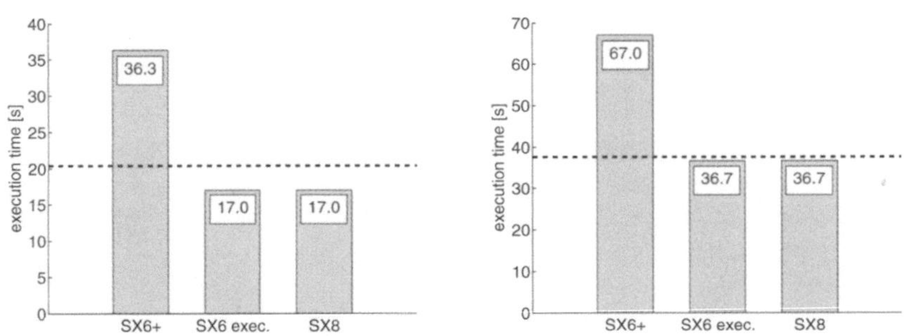

Fig. 10. Execution times on SX6+ and SX8 for an FCC crystal with 16k atoms using Grid Search (left) and IMD (right)

As Fig. 10 shows, our implementation of the Grid Search algorithm takes advantage of the new architectural features of the SX8. The speedup of 2.14 is noticeably larger than the expected 1.78. On the other hand, IMD stays in the expected range, with a speedup of 1.83. The annotation 'SX6 exec.' refers to times obtained with SX6 executables on the SX8.

Acknowledgements

The authors would like to thank Stefan Haberhauer for carrying out the VASP performance measurements.

References

1. F. Ercolessi, J. B. Adams, *Interatomic Potentials from First-Principles Calculations: the Force-Matching Method*, Europhys. Lett. **26** (1994) 583–588.
2. P. Brommer, F. Gähler, *Effective potentials for quasicrystals from ab-initio data*, Phil. Mag. **86** (2006) 753–758.
3. G. Kresse, J. Hafner, *Ab-initio molecular dynamics for liquid metals*, Phys. Rev. B **47** (1993) 558–561.
4. G. Kresse, J. Furthmüller, *Efficient iterative schemes for ab-initio total-energy calculations using a plane wave basis set*, Phys. Rev. B **54** (1996) 11169–11186.
5. G. Kresse, J. Furtmüller, *VASP – The Vienna Ab-initio Simulation Package*, `http://cms.mpi.univie.ac.at/vasp/`
6. J. Stadler, R. Mikulla, and H.-R. Trebin, *IMD: A Software Package for Molecular Dynamics Studies on Parallel Computers*, Int. J. Mod. Phys. C **8** (1997) 1131–1140 `http://www.itap.physk.uni-stuttgart.de/~imd`
7. M. S. Daw, M. I. Baskes, *Embedded-atom method: Derivation and application to impurities, surfaces, and other defects in metals*, Phys. Rev. B **29** (1984) 6443–6453.
8. G. S. Grest, B. Dünweg, K. Kremer, *Vectorized Link Cell Fortran Code for Molecular Dynamics Simulations for a Large Number of Particles*, Comp. Phys. Comm. **55** (1989) 269–285.
9. R. Everaers, K. Kremer, *A fast grid search algorithm for molecular dynamics simulations with short-range interactions*, Comp. Phys. Comm. **81** (1994) 19–55.

Molecular Simulation of Fluids with Short Range Potentials

Martin Bernreuther[1] and Jadran Vrabec[2]

[1] Institute of Parallel and Distributed Systems,
Simulation of Large Systems Department, University of Stuttgart,
Universitätsstraße 38, D-70569 Stuttgart, Germany,
martin.bernreuther@ipvs.uni-stuttgart.de,

[2] Institute of Thermodynamics and Thermal Process Engineering,
University of Stuttgart, Pfaffenwaldring 9, D-70569 Stuttgart, Germany,
vrabec@itt.uni-stuttgart.de

Abstract Molecular modeling and simulation of thermophysical properties using short-range potentials covers a large variety of real simple fluids and mixtures. To study nucleation phenomena within a research project, a molecular dynamics simulation package is developed. The target platform for this software are Clusters of Workstations (CoW), like the Linux cluster *Mozart* with 64 dual nodes, which is available at the Institute of Parallel and Distributed Systems, or the HLRS cluster *cacau*, which is part of the Teraflop Workbench. The used algorithms and data structures are discussed as well as first simulation results.

1 Physical and Mathematical Model

The Lennard-Jones (LJ) 12-6 potential [1]

$$u(r) = 4\varepsilon \left[\left(\frac{\sigma}{r}\right)^{12} - \left(\frac{\sigma}{r}\right)^{6} \right] \tag{1a}$$

is a semi-empiric function to describe the basic interactions between molecules. It covers both repulsion through the empiric r^{-12} term and dispersive attraction through the physically based r^{-6} term. Therefore it can be used to model the intermolecular interactions of non-polar or weakly polar fluids. In its simplest form, where only one Lennard-Jones site is present, it is well-suited for the simulation of inert gases and methane [2]. For molecular simulation programs, usually the dimensionless form is implemented

$$u^* = \frac{u}{\varepsilon} = 4 \left[\left(r^{*2}\right)^{-6} - \left(r^{*2}\right)^{-3} \right] \tag{1b}$$

with $r^* = r/\sigma$, where σ is the length parameter and ε is the energy parameter. In order to obtain a good description of the thermodynamic properties in most of the fluid region, which is of interest in the present work, they are preferably adjusted to experimental vapor-liquid equilibria [2]. Fluids consisting out of anisotropic molecules, can be modelled by composites of several LJ sites. When polar fluids are considered, additionally polar sites have to be added.

The molecular models in the present work are rigid and therefore have no internal degrees of freedom. To calculate the interactions between two multicentered molecules, all interactions between LJ centers are summed up.

Compared to phenomenological thermodynamic models, like equations of state or G^E-models, molecular models show superior predictive and extrapolative power. Furthermore, they allow a reliable and conceptually straightforward approach to the properties of fluid mixtures. In a binary mixture consisting of two components A and B, three different interactions are present: The two like interactions between molecules of the same component $A-A$ and $B-B$ and the unlike interaction between molecules of different kind $A-B$. In molecular simulation, usually pairwise additivity is assumed, so that the like interactions in a mixture are fully determined by the two pure substance models.

To determine the unlike Lennard-Jones parameters, the modified Lorentz-Berthelot combining rules provide a good starting point

$$\sigma_{AB} = \frac{\sigma_A + \sigma_B}{2} \tag{2a}$$

$$\varepsilon_{AB} = \xi\sqrt{\varepsilon_A \varepsilon_B} \tag{2b}$$

when the binary interaction parameter ξ is assumed to be unity. A refinement of the molecular model with respect to an accurate description of thermodynamic mixture properties can be achieved through an adjustment of ξ to one experimental bubble point of the mixture [3]. It has been shown for many mixtures, that ξ is typically within a 5% range around unity.

In molecular dynamics simulation, Newton's equations of motion are solved numerically for a number of N molecules over a period of time. These equations set up a system of ordinary differential equations of second order. This initial value problem can be solved with a time integration scheme like the Velocity-Störmer-Verlet method. During the simulation run the temperature is controlled with a thermostat to study the fluid at a specified state point. In the case of non-spherical molecules, an enhanced time integration procedure, which also takes care of orientation and angular velocity is needed [4].

2 Software Details

2.1 Existing Software

There are quite a few software packages for molecular dynamics simulations available on the internet. However, the ones we are aware of are all targeting different problem classes. The majority is made for biological applications with complex nonrigid molecules [5, 6, 7, 8]. There is also a powerful MD package for solid state physics [9], covering single site molecules only. The field of thermodynamics and process engineering is not visible here.

2.2 Framework

The present simulation package under development follows the classical preprocessing-calculation-postprocessing approach. A definition of the interfaces between these components is necessary. For this purpose a specific XML-based file format is used and allows a common data interchange (cf. Fig. 1). XML was chosen due to its flexibility. It is also a widespread standard [10] with broad support for many programming languages and numerous libraries are already available. But up to now XML is not a proper choice to store a large volume of binary data. Hence the phasespace, which contains the configuration (positions, velocities, orientations, angular velocities) for each molecule, and the molecule identifiers are stored in a binary file. To achieve platform independence and allow data interchange between machines of various architectures the "e**x**ternal **d**ata **r**epresentation" (XDR) standard [11] is used here. The main control file is a meta file, which contains the file name of the phasespace data file. It also contains the file names of XML-files to define the components used in the simulation. A large variety of these molecule type description files are kept in a directory as component library. The calculation engine not only gets its initial values from a given control file with its associated data files, it will also write these files in case of an interruption. Calculations may take a long time and a checkpointing facility

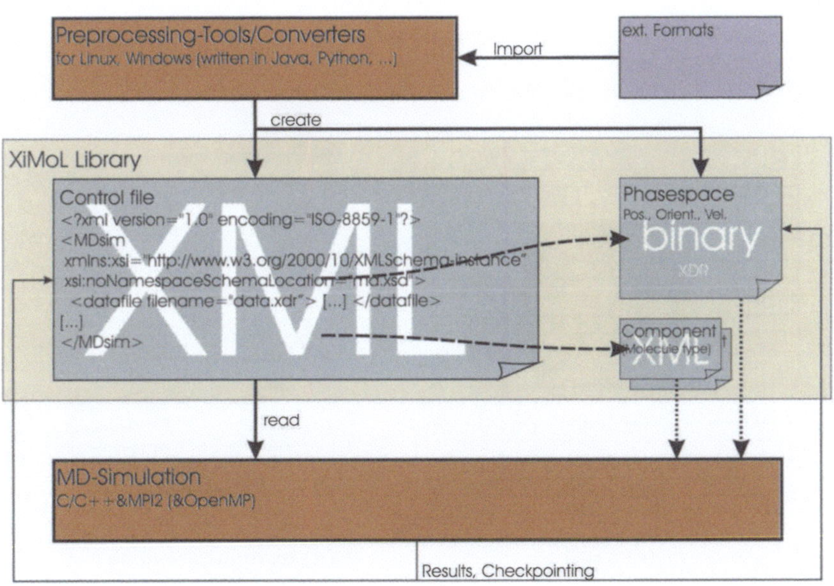

Fig. 1. Interfaces resp. IO within the framework

makes it possible to restart and continue the simulation run after an interruption. A library offers functions for reading and writing these files and thereby a common interface.

2.3 Algorithm and Data Structure

Assuming pairwise additivity, there are $\binom{N}{2} = \frac{1}{2}N(N-1)$ interactions for N molecules. Since LJ forces decay very fast with increasing distance (${r^*}^{-6}$), there are many small entries in the force matrix which may be replaced with zero for distances $r > r_c$. For this approximation the force matrix gets sparse with $\mathcal{O}(N)$ nonzero elements. The Linked-Cells algorithm gains a linear running time for these finite short-range potentials. The main idea is to decompose the domain

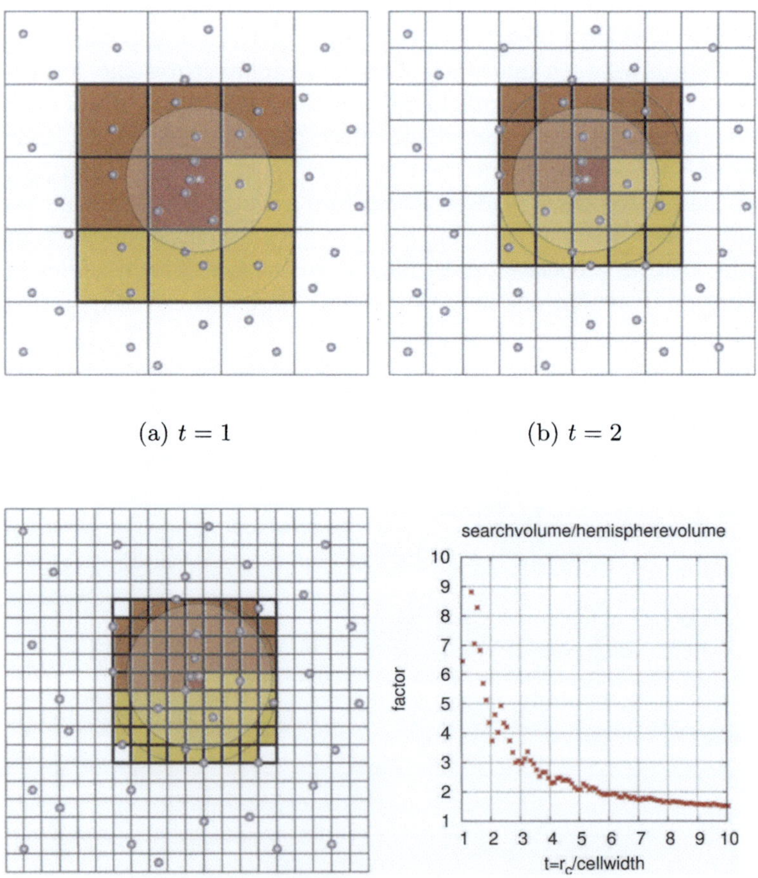

(a) $t = 1$ (b) $t = 2$

(c) $t = 4$ (d) examined volume

Fig. 2. Linked-Cells influence volume

into cuboid cells (cf. Fig. 2) and to assign molecules to the cells they are located in. The classical implementation uses cells of width r_c (cf. Fig. 2(a)). The cell influence volume is the union of all spheres with radius r_c whose centers are located inside the cell. This is a superset of the union of influence volumes for all molecules inside the cell. There is a direct volume representation of the influence volume, where the voxels correspond to the cells. This concept was generalized using cells of length r_c/t with $t \in \mathbb{R}^+$. The advantage is a higher flexibility and the possibility to increase the resolution. For $t \to \infty$ the examined volume will converge to the optimal Euclidean sphere and for $t \in \mathbb{N}^+$ a local optimum is obtained (cf. Fig. 2(d)). The data structure (cf. Fig. 3) is comparable to a hash table where a molecule location dependent hash function maps each molecule to an array entry and hash collisions are handled by lists. All atoms are additionally kept in a separate list (resp. one-dimensional array for the sequential version).

The drawback using this data structure with large t is the increasing runtime overhead, since a lot of empty cells have to be tested. In practice $t = 2$ is a good choice for fluid states [12, 13]. The implementation uses an one-dimensional array of pointers to molecules, which are heads of single linked intrusive lists. The domain is enlarged with a border "halo" region of width r_c, which takes care of the periodic boundary condition for a sequential version.

Neighbor cells are determined with the help of an offset vector (cf. Fig. 4(a)): the sum of the cell address and the offset leads to the neighbor cell address. The neighbor cell offsets are initialized once and cover only half of the cells influence

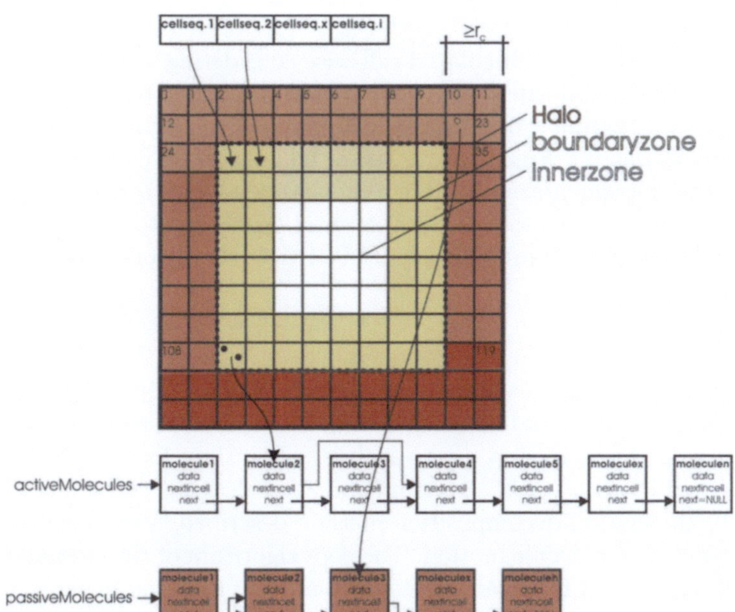

Fig. 3. Linked-Cells data structure

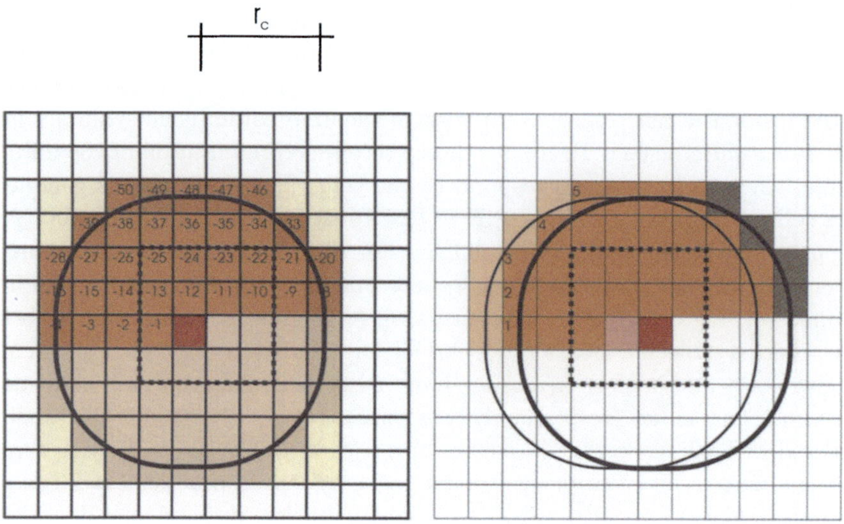

(a) offsets (b) moving to next cell

Fig. 4. Linked-Cells neighbors

volume to take advantage of Newton's third law (*actio* = *reactio*). As a result neighbor cells considered and left out within this region are point symmetric to the cell itself.

These interactions are calculated cell-wise considering the determined neighbor cells. The order influences the cache performance, due to a temporal locality of the data. Calculating a neighboring cell, most of the influence volume is part of the previous one (cf. Fig. 4(b)). A vector containing all cells to be considered simplifies the implementation of different strategies (e.g. applying space filling curves). The force calculation is the computationally most intensive part of the whole simulation with approximately 95% of the overall cost [14].

2.4 Parallelization

The target platform are clusters of workstations. Many installations use dual processor nodes, but shared memory and also hybrid parallelization will be done in a future step. The first step was to evaluate algorithms for distributed memory machines from literature, like the Atom and Force decomposition method [15]. In contrast to the Spatial decomposition method described later, both methods do not depend on the molecule motion. The core algorithm of the Atom decomposition (AD), also called Replicated data, is similar to a shared memory approach. Each processing element (PE) calculates the forces and new positions for one part of the molecules. All relevant data has to be provided and in the case of AD it has to be stored redundantly on each PE to be accessible. After each time step a synchronization of the redundant data is needed, which will inflate the

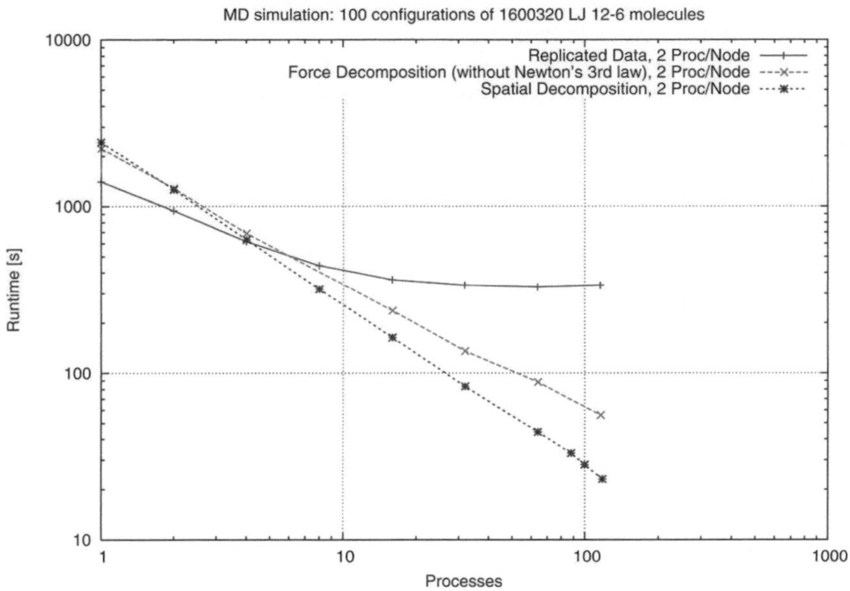

(a) runtime

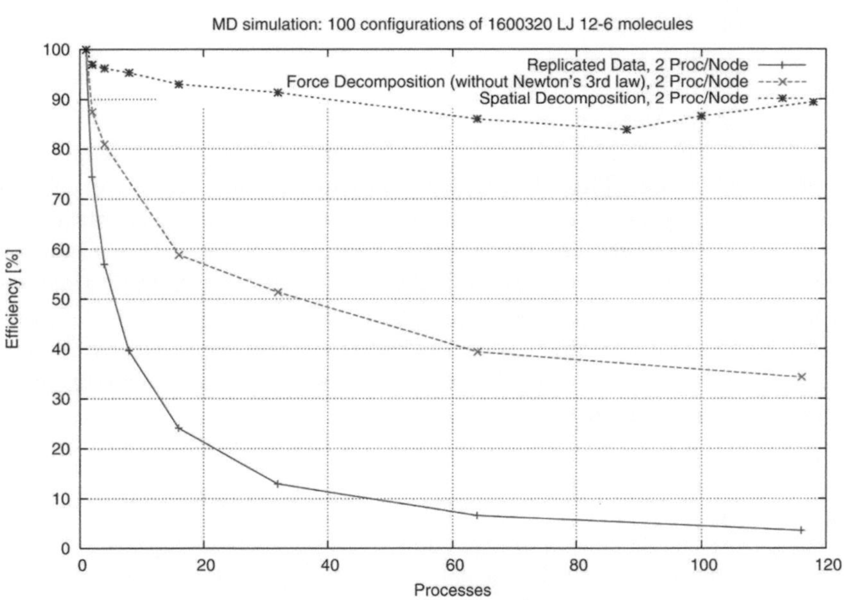

(b) efficiency

Fig. 5. Runtime results for parallel code on *Mozart*

communication effort. Regarding the Force decomposition (FD) method each PE is responsible for the calculation not only of a part of the molecule positions, but also of a block of the force matrix. A sophisticated reordering will result in an improved communication effort compared to the AD approach. The memory requirements are decreased in the same order. However, the number of PEs itself plays a role, e.g. prime numbers will result in force matrix slices for each PE and the FD will degenerate to an AD approach. The Spatial decomposition (SD) method will subdivide the domain with one subdomain for each PE. Each subdomain will have a cuboid shape here and will be placed in a cartesian topology. The PE needs access to data of neighbor PEs in the range of r_c. A "halo" region will accommodate copies of these molecules, which have to be synchronized. Since the halo region is approximately of lower dimension it only contains a relative small amount of molecules. Therefore the communication costs are less than the ones of the AD and FD method. Compared to these methods also the memory requirements for each PE are lower, which is of special interest for clusters with a large number of PEs with relatively small main memory like *cacau* (200 dual nodes each with 1 GB RAM for the majority of the nodes).

To make use of Newton's third law, additional communication is needed for all these methods, since the calculated force has to be transported to the associated PE. A recalculation might be faster, but this is dependent on the molecular model and its complexity. For simple single center molecules the SD method implemented uses a full "halo" (cf. Fig. 3) and doesn't make use of Newton's third law within the boundary region. Only the molecule positions of the "halo" molecules have to be communicated, which is done in 3 consecutive steps: first x, then y and final the z direction. The diagonal directions are done implicitly through multiple transportations.

Finally runtime tests on *Mozart* confirm the superiority of the SD method to the AD and FD method in terms of scalability (cf. Fig. 5) for homogeneous molecule distributions. This observation still can be made for the early stages of a nucleation process, but the picture changes later on, if large variations in the local densities occur. The latter is not the main focus of the actual work. Clusters with a dense package cause a higher computational effort for the related molecules and load balancing techniques have to be applied. In contrast to AD and FD methods, this is sophisticated for the SD method, especially for massively parallel systems, where the size of a subdomain is comparatively small. Further work will examine different strategies here.

3 Summary

Starting with the physical and mathematical background mainly details of an actual developed, emerging software project for the simulation of simple real fluids and prediction of thermodynamical properties were presented. The framework uses flexible XML metafiles combined with standardized binary files for the data exchange, which is extendable as well as scalable. The main component for the efficiency of the basic sequential algorithm is the Linked-Cells method with linear runtime complexity. The necessary parallelization for the CoW target platform is based on a spatial decomposition, which has proven to be superior to other known methods for the specific application area. The project is still at an early stage and good results obtained first on the development platform *Mozart* are also accomplishable on larger systems with similar architecture such as *cacau.*

Acknowledgements

This work is part of the project 688 "Massiv parallele molekulare Simulation und Visualisierung der Keimbildung in Mischungen für skalenübergreifende Modelle", which is financially supported by the Landesstiftung Baden-Württemberg within its program "Modellierung und Simulation auf Hochleistungscomputern".

References

1. M.P. Allen, D.J. Tildesley: Computer Simulation of Liquids. Oxford University Press, 2003 (reprint)
2. J. Vrabec, J. Stoll, H. Hasse: A set of molecular models for symmetric quadrupolar fluids. J. Phys. Chem. B **105** (2001) 12126–12133
3. J. Vrabec, J. Stoll, H. Hasse: Molecular models of unlike interactions in fluid mixtures. Molec. Sim. **31** (2005) 215–221
4. D. Fincham: Leapfrog rotational algorithms. Molec. Sim. **8** (1992) 165–178
5. Theoretical and Computational Biophysics Group, University of Illinois at Urbana-Champaign: NAMD. `http://www.ks.uiuc.edu/Research/namd/`
6. The Scripps Research Institute et al: Amber. `http://amber.scripps.edu/`
7. MD Group, University of Groningen: Gromacs. `http://www.gromacs.org/`
8. Laboratory for Computational Life Sciences, University of Notre Dame: Protomol. `http://www.nd.edu/ lcls/Protomol.html`
9. Institut für Theoretische und Angewandte Physik, Universität Stuttgart: IMD. `http://www.itap.physik.uni-stuttgart.de/ imd/`
10. World Wide Web Consortium: Extensible Markup Language (XML). `http://www.w3.org/XML/`
11. The Internet Engineering Task Force: XDR: External Data Representation Standard. `http://www.ietf.org/rfc/rfc1832.txt`
12. D. Mader: Molekulardynamische Simulation nanoskaliger Strömungsvorgänge. Master thesis, ITT, Universität Stuttgart, 2004
13. M. Bernreuther, H.-J. Bungartz: Molecular Simulation of Fluid Flow on a Cluster of Workstations. In: F. Hülsemann, M. Kowarschik, U. Rüde (ed.): 18th Symposium Simulationstechnique ASIM 2005 Proceedings, 2005
14. E. Miropolskiy: Implementation of parallel Algorithms for short-range molecular dynamics simulations. Student research project, IPVS, Universität Stuttgart, 2004
15. S. Plimpton: Fast parallel algorithms for short-range molecular dynamics. J. Comp. Phys. **117** (1995) 1–19

Toward TFlop Simulations of Supernovae

Konstantinos Kifonidis, Robert Buras, Andreas Marek, and Thomas Janka

Max Planck Institute for Astrophysics, Karl-Schwarzschild-Straße 1,
Postfach 1317, D-85741 Garching bei München, Germany,
kok@mpa-garching.mpg.de,
WWW home page: http://www.mpa-garching.mpg.de

Abstract We give an overview of the problems and the current status of (core collapse) supernova modelling, and report on our own recent progress, including the ongoing development of a code for multi-dimensional supernova simulations at TFlop speeds. In particular, we focus on the aspects of neutrino transport, and discuss the system of equations and the algorithm for its solution that are employed in this code. We also report first benchmark results from this code on an SGI Altix and a NEC SX-8.

1 Introduction

A star more massive than about 8 solar masses ends its live in a cataclysmic explosion, a supernova. Its quiescent evolution comes to an end, when the pressure in its inner layers is no longer able to balance the inward pull of gravity. Throughout its life, the star sustained this balance by generating energy through a sequence of nuclear fusion reactions, forming increasingly heavier elements in its core. However, when the core consists mainly of iron-group nuclei, central energy generation ceases. The fusion reactions producing iron-group nuclei relocate to the core's surface, and their "ashes" continuously increase the core's mass. Similar to a white dwarf, such a core is stabilized against gravity by the pressure of its degenerate gas of electrons. However, to remain stable, its mass must stay smaller than the Chandrasekhar limit. When the core grows larger than this limit, it collapses to a neutron star, and a huge amount ($\sim 10^{53}$ erg) of gravitational binding energy is set free. Most ($\sim 99\%$) of this energy is radiated away in neutrinos, but a small fraction is transferred to the outer stellar layers and drives the violent mass ejection which disrupts the star in a supernova.

Despite 40 years of research, the details of how this energy transfer happens and how the explosion is initiated are still not well understood. Observational evidence about the physical processes deep inside the collapsing star is sparse and almost exclusively indirect. The only direct observational access is via measurements of neutrinos or gravitational waves. To obtain insight into the events in the core, one must therefore heavily rely on sophisticated numerical simulations. The enormous amount of computer power required for this purpose has led to the use of several, often questionable, approximations and numerous ambiguous results in the past. Fortunately, however, the development of numerical tools and computational resources has meanwhile advanced to a point, where it is becoming possible to perform multi-dimensional simulations with unprecedented accuracy. Therefore there is hope that the physical processes which are essential for the explosion can finally be unraveled.

An understanding of the explosion mechanism is required to answer many important questions of nuclear, gravitational, and astro-physics like the following:

- How do the explosion energy, the explosion timescale, and the mass of the compact remnant depend on the progenitor's mass? Is the explosion mechanism the same for all progenitors? For which stars are black holes left behind as compact remnants instead of neutron stars?
- What is the role of rotation during the explosion? How rapidly do newly formed neutron stars rotate? What are the implications for gamma-ray burst ("collapsar") models?
- How do neutron stars receive their natal kicks? Are they accelerated by asymmetric mass ejection and/or anisotropic neutrino emission?
- How much Fe-group elements and radioactive isotopes (e.g., ^{22}Na, ^{44}Ti, ^{56,57}Ni) are produced during the explosion, how are these elements mixed into the mantle and envelope of the exploding star, and what does their observation tell us about the explosion mechanism? Are supernovae responsible for the production of very massive chemical elements by the so-called "rapid neutron capture process" or r-process?
- What are the generic properties of the neutrino emission and of the gravitational wave signal that are produced during stellar core collapse and explosion? Up to which distances could these signals be measured with operating or planned detectors on earth and in space? And what can one learn about supernova dynamics from a future measurement of such signals in case of a Galactic supernova?

2 Numerical Models

2.1 History and Constraints

According to theory, a shock wave is launched at the moment of "core bounce" when the neutron star begins to emerge from the collapsing stellar iron core. There is general agreement, supported by all "modern" numerical simulations, that this shock is unable to propagate directly into the stellar mantle and envelope, because it looses too much energy in dissociating iron into free nucleons while it moves through the outer core. The "prompt" shock ultimately stalls. Thus the currently favored theoretical paradigm makes use of the fact that a huge energy reservoir is present in the form of neutrinos, which are abundantly emitted from the hot, nascent neutron star. The absorption of electron neutrinos and antineutrinos by free nucleons in the post shock layer is thought to reenergize the shock, and lead to the supernova explosion.

Detailed *spherically symmetric* hydrodynamic models, which recently include a very accurate treatment of the time-dependent, multi-flavor, multi-frequency neutrino transport based on a numerical solution of the Boltzmann transport equation [1, 2, 3, 4], reveal that this "delayed, neutrino-driven mechanism" does not work as simply as originally envisioned. Although in principle able to trigger the explosion (e.g., [5], [6], [7]), neutrino energy transfer to the postshock matter turned out to be too weak. For inverting the infall of the stellar core and initiating powerful mass ejection, an increase of the efficiency of neutrino energy deposition is needed.

A number of physical phenomena have been pointed out that can enhance neutrino energy deposition behind the stalled supernova shock. They are all linked to the fact that the real world is multi-dimensional instead of spherically symmetric (or one-dimensional; 1D) as assumed in the work cited above:

(1) Convective instabilities in the neutrino-heated layer between the neutron star and the supernova shock develop to violent convective overturn [8]. This convective overturn is helpful for the explosion, mainly because (a) neutrino-heated matter rises and increases the pressure behind the shock, thus pushing the shock further out, and (b) cool matter is able to penetrate closer to the neutron star where it can absorb neutrino energy more efficiently. Both effects allow multi-dimensional models to explode easier than spherically symmetric ones [9, 10, 11].
(2) Recent work [12, 13, 14, 15] has demonstrated that the stalled supernova shock is also subject to a second non-radial instability which can grow to a dipolar, global deformation of the shock [15].
(3) Convective energy transport inside the nascent neutron star [16, 17, 18] might enhance the energy transport to the neutrinosphere and could thus boost the neutrino luminosities. This would in turn increase the neutrino-heating behind the shock.
(4) Rapid rotation of the collapsing stellar core and of the neutron star could lead to direction-dependent neutrino emission [19, 20] and thus anisotropic neutrino heating [21, 22]. Centrifugal forces, meridional circulation, pole-to-equator differences of the stellar structure, and magnetic fields could also have important consequences for the supernova evolution.

This list of multi-dimensional phenomena awaits more detailed exploration in multi-dimensional simulations. Until recently, such simulations have been performed with only a grossly simplified treatment of the involved microphysics, in particular of the neutrino transport and neutrino-matter interactions. At best, grey (i.e., single energy) flux-limited diffusion schemes were employed. All published successful simulations of supernova explosions by the convectively aided neutrino-heating mechanism in two [9, 10, 23, 24] and three dimensions [25, 26] used such a radical approximation of the neutrino transport.

Since, however, the role of the neutrinos is crucial for the problem, and because previous experience shows that the outcome of simulations is indeed very sensitive to the employed transport approximations, studies of the explosion mechanism require the best available description of the neutrino physics. This implies that one has to solve the Boltzmann transport equation for neutrinos.

2.2 Recent Calculations and the Need for TFlop Simulations

We have recently advanced to a new level of accuracy for supernova simulations by generalizing the VERTEX code, a Boltzmann solver for neutrino transport, from spherical symmetry [27] to multi-dimensional applications [28, 29, 30]. The corresponding mathematical model, and in particular our method for tackling the integro-differential transport problem in multi-dimensions, will be summarized in Sect. 3.

Results of a set of simulations with our code in 1D and 2D for progenitor stars with different masses have recently been published by [28], and with respect to the expected gravitational-wave signals from rotating and convective supernova cores by [31]. The recent progress in supernova modeling was summarized and set in perspective in a conference article by [29].

Our collection of simulations has helped us to identify a number of effects which have brought our two-dimensional models close to the threshold of explosion. This makes us optimistic that the solution of the long-standing problem of how massive stars explode may be in reach. In particular, we have recognized the following aspects as advantageous:

- Stellar rotation, even at a moderate level, supports the expansion of the stalled shock by centrifugal forces and instigates overturn motion in the neutrino-heated postshock matter by meridional circulation flows in addition to convective instabilities.
- Changing from the current "standard" and most widely used equation of state (EoS) for stellar core-collapse simulations [32] to alternative descriptions [33, 34], we found in 1D calculations that a higher incompressibility of the supranuclear phase yields a less dramatic and less rapid recession of the stalled shock after it has reached its maximum expansion [35]. This finding suggests that the EoS of [34] might lead to more favorable conditions for strong postshock convection, and thus more efficient neutrino heating, than current 2D simulations with the EoS of [32].
- Enlarging the two-dimensional grid from a 90° to a full 180° wedge, we indeed discovered global dipolar shock oscillations and a strong tendency for the growth of $l = 1, 2$ modes as observed also in previous models with a simplified treatment of neutrino transport [15]. The dominance of low-mode convection helped the expansion of the supernova shock in the 180°-simulation of an $11.2\,M_\odot$ star. In fact, the strongly deformed shock had expanded to a radius of more than 600 km at 226 ms post bounce with no tendency to return (Fig. 1). This model was on the way to an explosion, although probably a weak one, in contrast to simulations of the same star with a constrained

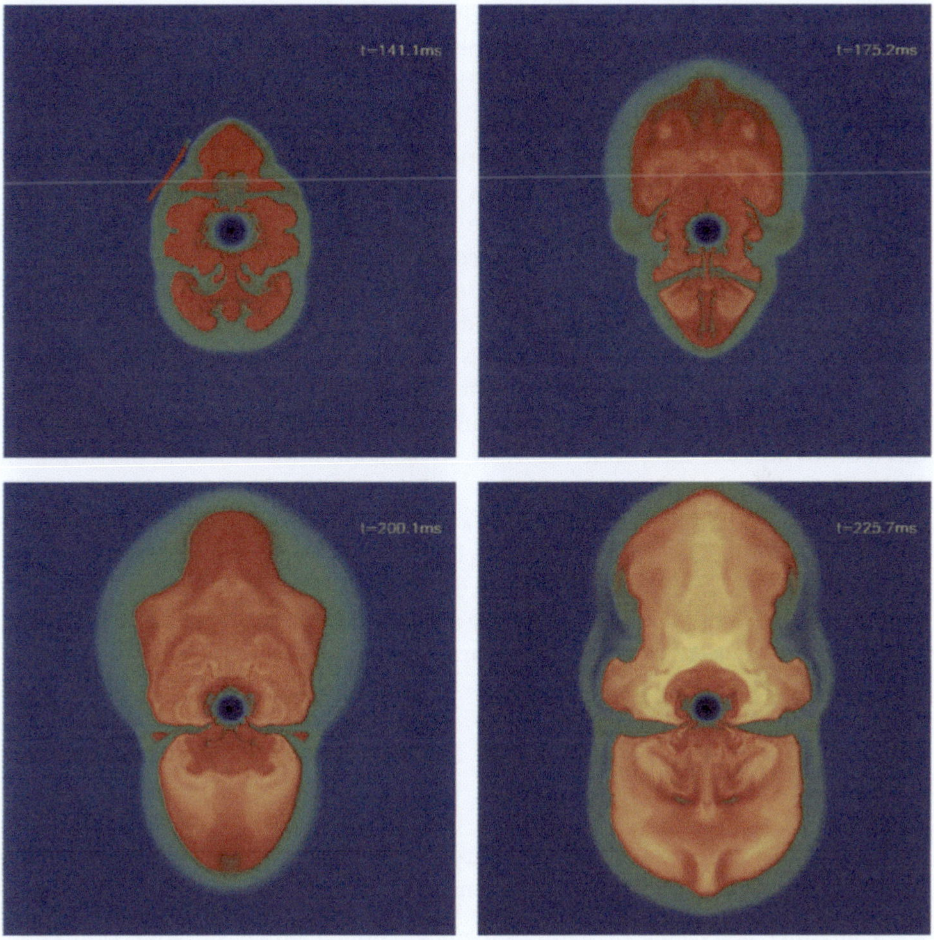

Fig. 1. Sequence of snapshots showing the large-scale convective overturn in the neutrino-heated postshock layer at four post-bounce times ($t_{\mathrm{pb}} = 141.1\,\mathrm{ms}$, $175.2\,\mathrm{ms}$, $200.1\,\mathrm{ms}$, and $225.7\,\mathrm{ms}$, from top left to bottom right) during the evolution of a (non-rotating) $11.2\,M_\odot$ progenitor star. The entropy is color coded with highest values being represented by red and yellow, and lowest values by blue and black. The dense neutron star is visible as a low-entropy circle at the center. A convective layer interior to the neutrinosphere cannot be visualized with the employed color scale because the entropy contrast there is small. Convection in this layer is driven by a negative gradient of the lepton number. The computation was performed with spherical coordinates, assuming axial symmetry, and employing the "ray-by-ray plus" variable Eddington factor technique for treating neutrino transport in multi-dimensional supernova simulations. Equatorial symmetry is broken on large scales soon after bounce, and low-mode convection begins to dominate the flow between the neutron star and the strongly deformed supernova shock. The model continues to develop a weak explosion. The scale of the plots is 1200 km in both directions.

90° wedge [29]. Unfortunately, calculating the first 226 ms of the evolution of this model already required about half a year of computer time on a 32 processor IBM p690, so that we were not able to continue the simulation to still later post-bounce times.

All these effects are potentially important, and some (or even all of them) may represent crucial ingredients for a successful supernova simulation. So far no multi-dimensional calculations have been performed, in which two or more of these items have been taken into account simultaneously, and thus their mutual interaction awaits to be investigated. It should also be kept in mind that our knowledge of supernova microphysics, and especially the EoS of neutron star matter, is still incomplete, which implies major uncertainties for supernova modeling. Unfortunately, the impact of different descriptions for this input physics has so far not been satisfactorily explored with respect to the neutrino-heating mechanism and the long-time behavior of the supernova shock, in particular in multi-dimensional models.

From this it is clear that rather extensive parameter studies using multi-dimensional simulations are required to identify the physical processes which are essential for the explosion. Since on a dedicated machine performing at a sustained speed of about 30 GFlops already a single 2D simulation has a turn-around time of more than half a year, these parameter studies are not possible without TFlop simulations.

3 The Mathematical Model

The non-linear system of partial differential equations which is solved in our code consists of the following components:

- The Euler equations of hydrodynamics, supplemented by advection equations for the electron fraction and the chemical composition of the fluid, and formulated in spherical coordinates;
- the Poisson equation for calculating the gravitational source terms which enter the Euler equations, including corrections for general relativistic effects;
- the Boltzmann transport equation which determines the (non-equilibrium) distribution function of the neutrinos;
- the emission, absorption, and scattering rates of neutrinos, which are required for the solution of the Boltzmann equation;
- the equation of state of the stellar fluid, which provides the closure relation between the variables entering the Euler equations, i.e. density, momentum, energy, electron fraction, composition, and pressure.

In what follows we will briefly summarize the neutrino transport algorithms. For a more complete description of the entire code we refer the reader to [28], [30], and the references therein.

3.1 "Ray-by-ray plus" Variable Eddington Factor Solution of the Neutrino Transport Problem

The crucial quantity required to determine the source terms for the energy, momentum, and electron fraction of the fluid owing to its interaction with the neutrinos is the neutrino distribution function in phase space, $f(r, \vartheta, \phi, \epsilon, \Theta, \Phi, t)$. Equivalently, the neutrino intensity $I = c/(2\pi\hbar c)^3 \cdot \epsilon^3 f$ may be used. Both are seven-dimensional functions, as they describe, at every point in space (r, ϑ, ϕ), the distribution of neutrinos propagating with energy ϵ into the direction (Θ, Φ) at time t (Fig. 2).

The evolution of I (or f) in time is governed by the Boltzmann equation, and solving this equation is, in general, a six-dimensional problem (since time is usually not counted as a separate dimension). A solution of this equation by direct discretization (using an S_N scheme) would require computational resources in the PetaFlop range. Although there are attempts by at least one group in the United States to follow such an approach, we feel that, with the currently available computational resources, it is mandatory to reduce the dimensionality of the problem.

Actually this should be possible, since the source terms entering the hydrodynamic equations are *integrals* of I over momentum space (i.e. over ϵ, Θ, and Φ), and thus only a fraction of the information contained in I is truly required to compute the dynamics of the flow. It makes therefore sense to consider angular moments of I, and to solve evolution equations for these moments, instead of dealing with the Boltzmann equation directly. The 0th to 3rd order moments are defined as

$$J, \boldsymbol{H}, \mathbf{K}, \mathbf{L}, \ldots (r, \vartheta, \phi, \epsilon, t) = \frac{1}{4\pi} \int I(r, \vartheta, \phi, \epsilon, \Theta, \Phi, t)\, \boldsymbol{n}^{0,1,2,3,\ldots}\, \mathrm{d}\Omega \qquad (1)$$

where $\mathrm{d}\Omega = \sin\Theta\, \mathrm{d}\Theta\, \mathrm{d}\Phi$, $\boldsymbol{n} = (\sin\Theta\cos\Phi, \sin\Theta\sin\Phi, \cos\Theta)$, and exponentiation represents repeated application of the dyadic product. Note that the moments are *tensors* of the required rank.

This leaves us with a four-dimensional problem. So far no approximations have been made. In order to reduce the size of the problem even further, one

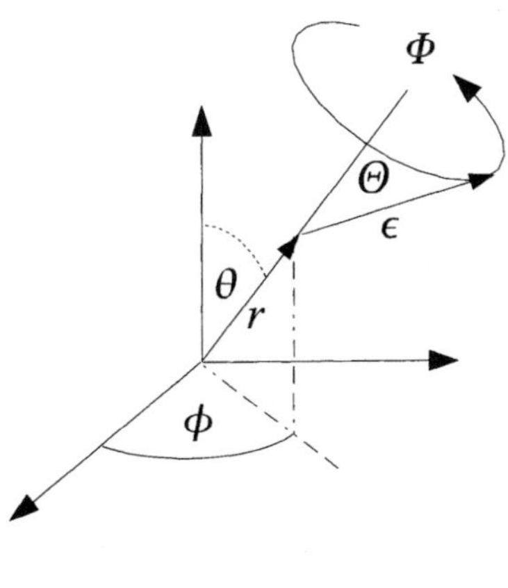

Fig. 2. Illustration of the phase space coordinates (see the main text).

needs to resort to assumptions on its symmetry. At this point, one usually employs azimuthal symmetry for the stellar matter distribution, i.e. any dependence on the azimuth angle ϕ is ignored, which implies that the hydrodynamics of the problem can be treated in two dimensions. It also implies $I(r,\vartheta,\epsilon,\Theta,\Phi) = I(r,\vartheta,\epsilon,\Theta,-\Phi)$. If, in addition, it is assumed that I is even independent of Φ, then each of the angular moments of I becomes a *scalar*, which depends on two spatial dimensions, and one dimension in momentum space: $J,H,K,L = J,H,K,L(r,\vartheta,\epsilon,t)$. Thus we have reduced the problem to three dimensions in total.

The System of Equations

With the aforementioned assumptions it can be shown [30], that in order to compute the source terms for the energy and electron fraction of the fluid, the following two transport equations need to be solved:

$$\begin{aligned}
&\left(\frac{1}{c}\frac{\partial}{\partial t} + \beta_r\frac{\partial}{\partial r} + \frac{\boldsymbol{\beta_\vartheta}}{\boldsymbol{r}}\frac{\boldsymbol{\partial}}{\boldsymbol{\partial\vartheta}}\right) J + J\left(\frac{1}{r^2}\frac{\partial(r^2\beta_r)}{\partial r} + \frac{1}{\boldsymbol{r\sin\vartheta}}\frac{\boldsymbol{\partial(\sin\vartheta\beta_\vartheta)}}{\boldsymbol{\partial\vartheta}}\right)\\
&+\frac{1}{r^2}\frac{\partial(r^2H)}{\partial r} + \frac{\beta_r}{c}\frac{\partial H}{\partial t} - \frac{\partial}{\partial\epsilon}\left\{\frac{\epsilon}{c}\frac{\partial\beta_r}{\partial t}H\right\} - \frac{\partial}{\partial\epsilon}\left\{\epsilon J\left(\frac{\beta_r}{r} + \frac{1}{\boldsymbol{2r\sin\vartheta}}\frac{\boldsymbol{\partial(\sin\vartheta\beta_\vartheta)}}{\boldsymbol{\partial\vartheta}}\right)\right\}\\
&-\frac{\partial}{\partial\epsilon}\left\{\epsilon K\left(\frac{\partial\beta_r}{\partial r} - \frac{\beta_r}{r} - \frac{1}{\boldsymbol{2r\sin\vartheta}}\frac{\boldsymbol{\partial(\sin\vartheta\beta_\vartheta)}}{\boldsymbol{\partial\vartheta}}\right)\right\} + J\left(\frac{\beta_r}{r} + \frac{1}{\boldsymbol{2r\sin\vartheta}}\frac{\boldsymbol{\partial(\sin\vartheta\beta_\vartheta)}}{\boldsymbol{\partial\vartheta}}\right)\\
&\qquad + K\left(\frac{\partial\beta_r}{\partial r} - \frac{\beta_r}{r} - \frac{1}{\boldsymbol{2r\sin\vartheta}}\frac{\boldsymbol{\partial(\sin\vartheta\beta_\vartheta)}}{\boldsymbol{\partial\vartheta}}\right) + \frac{2}{c}\frac{\partial\beta_r}{\partial t}H = C^{(0)}, \qquad (2)
\end{aligned}$$

$$\begin{aligned}
&\left(\frac{1}{c}\frac{\partial}{\partial t} + \beta_r\frac{\partial}{\partial r} + \frac{\boldsymbol{\beta_\vartheta}}{\boldsymbol{r}}\frac{\boldsymbol{\partial}}{\boldsymbol{\partial\vartheta}}\right) H + H\left(\frac{1}{r^2}\frac{\partial(r^2\beta_r)}{\partial r} + \frac{1}{\boldsymbol{r\sin\vartheta}}\frac{\boldsymbol{\partial(\sin\vartheta\beta_\vartheta)}}{\boldsymbol{\partial\vartheta}}\right)\\
&\qquad + \frac{\partial K}{\partial r} + \frac{3K-J}{r} + H\left(\frac{\partial\beta_r}{\partial r}\right) + \frac{\beta_r}{c}\frac{\partial K}{\partial t} - \frac{\partial}{\partial\epsilon}\left\{\frac{\epsilon}{c}\frac{\partial\beta_r}{\partial t}K\right\}\\
&\qquad - \frac{\partial}{\partial\epsilon}\left\{\epsilon L\left(\frac{\partial\beta_r}{\partial r} - \frac{\beta_r}{r} - \frac{1}{\boldsymbol{2r\sin\vartheta}}\frac{\boldsymbol{\partial(\sin\vartheta\beta_\vartheta)}}{\boldsymbol{\partial\vartheta}}\right)\right\}\\
&\quad - \frac{\partial}{\partial\epsilon}\left\{\epsilon H\left(\frac{\beta_r}{r} + \frac{1}{\boldsymbol{2r\sin\vartheta}}\frac{\boldsymbol{\partial(\sin\vartheta\beta_\vartheta)}}{\boldsymbol{\partial\vartheta}}\right)\right\} + \frac{1}{c}\frac{\partial\beta_r}{\partial t}(J+K) = C^{(1)}. \qquad (3)
\end{aligned}$$

These are evolution equations for the neutrino energy density, J, and the neutrino flux, H, and follow from the zeroth and first moment equations of the comoving frame (Boltzmann) transport equation in the Newtonian, $\mathcal{O}(v/c)$ approximation. The quantities $C^{(0)}(J,H)$ and $C^{(1)}(J,H)$ are source terms that result from the collision term of the Boltzmann equation, while $\beta_r = v_r/c$ and $\beta_\vartheta = v_\vartheta/c$, where v_r and v_ϑ are the components of the hydrodynamic velocity, and c is the speed of light. The functional dependences $\beta_r = \beta_r(r,\vartheta,t)$, $J = J(r,\vartheta,\epsilon,t)$, etc. are suppressed in the notation. This system includes four unknown moments (J,H,K,L) but only two equations, and thus needs to be supplemented by two more relations. This is done by substituting $K = f_K \cdot J$ and $L = f_L \cdot J$, where f_K and f_L are the variable Eddington factors, which for the moment may be regarded as being known, but in general must be determined from a separate system of equations (see below).

A finite volume discretization of Eqs. (2–3) is sufficient to guarantee exact conservation of the total neutrino energy. However, and as described in detail in [27], it is not sufficient to guarantee also exact conservation of the neutrino number. To achieve this, we discretize and solve a set of two additional equations. With $\mathcal{J} = J/\epsilon$, $\mathcal{H} = H/\epsilon$, $\mathcal{K} = K/\epsilon$, and $\mathcal{L} = L/\epsilon$, this set of equations reads

$$\left(\frac{1}{c}\frac{\partial}{\partial t} + \beta_r \frac{\partial}{\partial r} + \boldsymbol{\frac{\beta_\vartheta}{r}\frac{\partial}{\partial \vartheta}}\right)\mathcal{J} + \mathcal{J}\left(\frac{1}{r^2}\frac{\partial(r^2\beta_r)}{\partial r} + \boldsymbol{\frac{1}{r\sin\vartheta}\frac{\partial(\sin\vartheta\beta_\vartheta)}{\partial\vartheta}}\right)$$
$$+\frac{1}{r^2}\frac{\partial(r^2\mathcal{H})}{\partial r} + \frac{\beta_r}{c}\frac{\partial \mathcal{H}}{\partial t} - \frac{\partial}{\partial\epsilon}\left\{\frac{\epsilon}{c}\frac{\partial\beta_r}{\partial t}\mathcal{H}\right\} - \frac{\partial}{\partial\epsilon}\left\{\epsilon\mathcal{J}\left(\frac{\beta_r}{r} + \boldsymbol{\frac{1}{2r\sin\vartheta}\frac{\partial(\sin\vartheta\beta_\vartheta)}{\partial\vartheta}}\right)\right\}$$
$$-\frac{\partial}{\partial\epsilon}\left\{\epsilon\mathcal{K}\left(\frac{\partial\beta_r}{\partial r} - \frac{\beta_r}{r} - \boldsymbol{\frac{1}{2r\sin\vartheta}\frac{\partial(\sin\vartheta\beta_\vartheta)}{\partial\vartheta}}\right)\right\} + \frac{1}{c}\frac{\partial\beta_r}{\partial t}\mathcal{H} = \mathcal{C}^{(0)}, \quad (4)$$

$$\left(\frac{1}{c}\frac{\partial}{\partial t} + \beta_r \frac{\partial}{\partial r} + \boldsymbol{\frac{\beta_\vartheta}{r}\frac{\partial}{\partial \vartheta}}\right)\mathcal{H} + \mathcal{H}\left(\frac{1}{r^2}\frac{\partial(r^2\beta_r)}{\partial r} + \boldsymbol{\frac{1}{r\sin\vartheta}\frac{\partial(\sin\vartheta\beta_\vartheta)}{\partial\vartheta}}\right)$$
$$+\frac{\partial\mathcal{K}}{\partial r} + \frac{3\mathcal{K} - \mathcal{J}}{r} + \mathcal{H}\left(\frac{\partial\beta_r}{\partial r}\right) + \frac{\beta_r}{c}\frac{\partial\mathcal{K}}{\partial t} - \frac{\partial}{\partial\epsilon}\left\{\frac{\epsilon}{c}\frac{\partial\beta_r}{\partial t}\mathcal{K}\right\}$$
$$-\frac{\partial}{\partial\epsilon}\left\{\epsilon\mathcal{L}\left(\frac{\partial\beta_r}{\partial r} - \frac{\beta_r}{r} - \boldsymbol{\frac{1}{2r\sin\vartheta}\frac{\partial(\sin\vartheta\beta_\vartheta)}{\partial\vartheta}}\right)\right\}$$
$$-\frac{\partial}{\partial\epsilon}\left\{\epsilon\mathcal{H}\left(\frac{\beta_r}{r} + \boldsymbol{\frac{1}{2r\sin\vartheta}\frac{\partial(\sin\vartheta\beta_\vartheta)}{\partial\vartheta}}\right)\right\} - \mathcal{L}\left(\frac{\partial\beta_r}{\partial r} - \frac{\beta_r}{r} - \boldsymbol{\frac{1}{2r\sin\vartheta}\frac{\partial(\sin\vartheta\beta_\vartheta)}{\partial\vartheta}}\right)$$
$$-\mathcal{H}\left(\frac{\beta_r}{r} + \boldsymbol{\frac{1}{2r\sin\vartheta}\frac{\partial(\sin\vartheta\beta_\vartheta)}{\partial\vartheta}}\right) + \frac{1}{c}\frac{\partial\beta_r}{\partial t}\mathcal{J} = \mathcal{C}^{(1)}. \quad (5)$$

The moment Eqs. (2–5) are very similar to the $\mathcal{O}(v/c)$ equations in spherical symmetry which were solved in the 1D simulations of [27] (see Eqs. (7), (8), (30), and (31) of the latter work). This similarity has allowed us to reuse a good fraction of the one-dimensional version of VERTEX, for coding the multi-dimensional algorithm. The additional terms necessary for this purpose have been set in boldface above.

Finally, the changes of the energy, e, and electron fraction, Y_{e}, required for the hydrodynamics are given by the following two equations

$$\frac{\mathrm{d}e}{\mathrm{d}t} = -\frac{4\pi}{\rho}\int_0^\infty \mathrm{d}\epsilon \sum_{\nu\in(\nu_{\mathrm{e}},\bar{\nu}_{\mathrm{e}},\dots)} C_\nu^{(0)}\left(J(\epsilon), H(\epsilon)\right), \quad (6)$$

$$\frac{\mathrm{d}Y_{\mathrm{e}}}{\mathrm{d}t} = -\frac{4\pi\, m_{\mathrm{B}}}{\rho}\int_0^\infty \mathrm{d}\epsilon \left(\mathcal{C}_{\nu_{\mathrm{e}}}^{(0)}\left(\mathcal{J}(\epsilon), \mathcal{H}(\epsilon)\right) - \mathcal{C}_{\bar{\nu}_{\mathrm{e}}}^{(0)}\left(\mathcal{J}(\epsilon), \mathcal{H}(\epsilon)\right)\right) \quad (7)$$

(for the momentum source terms due to neutrinos see [30]). Here m_{B} is the baryon mass, and the sum in Eq. (6) runs over all neutrino types. The full system consisting of Eqs. (2–7) is stiff, and thus requires an appropriate discretization scheme for its stable solution.

Method of Solution

In order to discretize Eqs. (2–7), the spatial domain $[0, r_{\max}] \times [\vartheta_{\min}, \vartheta_{\max}]$ is covered by N_r radial, and N_ϑ angular zones, where $\vartheta_{\min} = 0$ and $\vartheta_{\max} = \pi$ correspond to the north and south poles, respectively, of the spherical grid. (In general, we allow for grids with different radial resolutions in the neutrino transport and hydrodynamic parts of the code. The number of radial zones for the hydrodynamics will be denoted by N_r^{hyd}.) The number of bins used in energy space is N_ϵ and the number of neutrino types taken into account is N_ν.

The equations are solved in two operator-split steps corresponding to a lateral and a radial sweep.

In the first step, we treat the boldface terms in the respectively first lines of Eqs. (2–5), which describe the lateral advection of the neutrinos with the stellar fluid, and thus couple the angular moments of the neutrino distribution of neighbouring angular zones. For this purpose we consider the equation

$$\frac{1}{c}\frac{\partial \Xi}{\partial t} + \frac{1}{r \sin\vartheta} \frac{\partial(\sin\vartheta \, \beta_\vartheta \, \Xi)}{\partial \vartheta} = 0\,, \tag{8}$$

where Ξ represents one of the moments J, H, $\mathcal{J}$, or $\mathcal{H}$. Although it has been suppressed in the above notation, an equation of this form has to be solved for each radius, for each energy bin, and for each type of neutrino. An explicit upwind scheme is used for this purpose.

In the second step, the radial sweep is performed. Several points need to be noted here:

- terms in boldface not yet taken into account in the lateral sweep, need to be included into the discretization scheme of the radial sweep. This can be done in a straightforward way since these remaining terms do not include ϑ-derivatives of the transport variables (J, H) or $(\mathcal{J}, \mathcal{H})$. They only include ϑ-derivatives of the hydrodynamic velocity v_ϑ, which is a *constant* scalar field for the transport problem.
- the right hand sides (source terms) of the equations and the coupling in energy space have to be accounted for. The coupling in energy is non-local, since the source terms of Eqs. (2–5) stem from the Boltzmann equation, which is an integro-differential equation and couples all the energy bins
- the discretization scheme for the radial sweep is *implicit* in time. Explicit schemes would require very small time steps to cope with the stiffness of the source terms in the optically thick regime, and the small CFL time step dictated by neutrino propagation with the speed of light in the optically thin regime. Still, even with an implicit scheme $\gtrsim 10^5$ time steps are required per simulation. This makes the calculations expensive.

Once the equations for the radial sweep have been discretized in radius and energy, the resulting solver is applied ray-by-ray for each angle ϑ and for each type of neutrino, i.e. for constant ϑ, N_ν two-dimensional problems need to be solved.

The discretization itself is done using a second order accurate scheme with backward differencing in time according to [27]. This leads to a non-linear system of algebraic equations, which is solved by Newton-Raphson iteration with explicit construction and inversion of the corresponding Jacobian matrix.

Inversion of the Jacobians

The Jacobians resulting from the radial sweep are block-pentadiagonal matrices with $2 \times N_r + 1$ rows of blocks. The blocks themselves are dense, because of the non-local coupling in energy. For the transport of electron neutrinos and antineutrinos, the blocks are of dimension $(2 \times N_\epsilon + 2)^2$, or $(4 \times N_\epsilon + 2)^2$, depending, respectively, on whether only Eqs. (2), (3), (6), and (7) or the full system consisting of Eqs. (2–7) is solved (see below).

Three alternative direct methods are currently implemented for solving the resulting linear systems. The first is a Block-Thomas solver which uses optimized routines from the BLAS and LAPACK libraries to perform the necessary LU decompositions and backsubstitutions of the dense blocks. In this case vectorization is used within BLAS, i.e. within the operations on blocks, and the achievable vector length is determined by the block size.

The second is a block cyclic reduction solver which also uses BLAS and LAPACK for block operations.

The third is a block cyclic reduction solver that is vectorized along the Jacobians' diagonals (i.e. along the radius, r) in order to obtain longer vector lengths. This might be of advantage in case a simulation needs to be set up with a small resolution in energy space, resulting in a correspondingly small size of the single blocks.

Variable Eddington Factors

To solve Eqs. (2–7), we need the variable Eddington factors $f_K = K/J$ and $f_L = L/J$. These closure relations are obtained from the solution of a simplified ("model") Boltzmann equation. The integro-differential character of this equation is tackled by expressing the angular integrals in the interaction kernels of its right-hand side, with the moments J and H, for which estimates are obtained from a solution of the system of moment equations (2–3), (6) and (7). With the right-hand side known, the model Boltzmann equation is solved by means of the so-called tangent ray method (see [36], and [27] for details), and the entire procedure is iterated until convergence of the Eddington factors is achieved (cf. Fig. 3).

Note that this apparently involved procedure is computationally efficient, because the Eddington factors are geometrical quantities, which vary only slowly, and thus can be computed relatively cheaply using only a "model" transport equation. Note also that only the system of Eqs. (2–3), (6) and (7), and not the full system Eqs. (2–7), is used in the iteration. This allows us to save computer

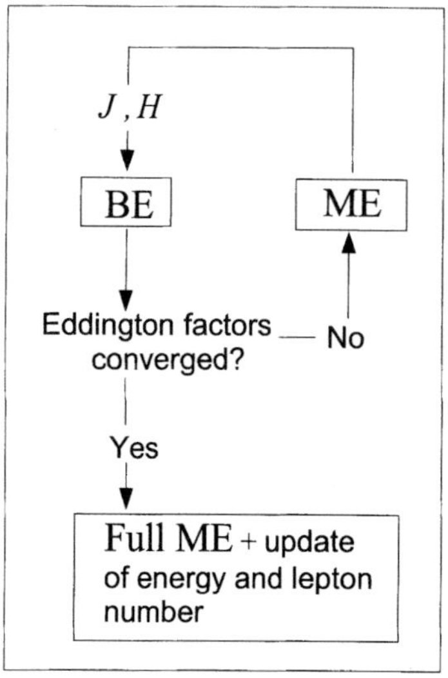

Fig. 3. Illustration of the iteration procedure for calculating the variable Eddington factors. The boxes labeled ME and BE represent the solution algorithms for the moment equations, and the "model" Boltzmann equation, respectively (see the text for details).

time. Once the Eddington factors are known, the complete system Eqs. (2–7), enforcing conservation of energy and neutrino number, is solved once, in order to update the energy and electron fraction (lepton number) of the fluid.

In contrast to previous work [27, 30], our latest code version takes into account that the Eddington factors are functions of radius and angle, $f_K = f_K(r, \vartheta, t)$ and $f_L = f_L(r, \vartheta, t)$, and thus the iteration procedure shown in Fig. 3 is applied on each ray, i.e. for each ϑ.

4 Implementation and First Benchmarks

The VERTEX routines, that have been described above, have been coupled to the hydrodynamics code PROMETHEUS, to obtain the full supernova code PROMETHEUS/VERTEX. In a typical low-resolution supernova simulation, like the one shown in Fig. 1 and corresponding to setup "S" of Table 1, the VERTEX transport routines typically account for 99.5%, and the hydrodynamics for about 0.5% of the entire execution time. The ratio of computing times is expected to tilt even further towards the transport side when the larger setups in Table 1 are investigated (especially the one with 34 energy bins), since a good fraction of the total time is spent in inverting the Jacobians. It is thus imperative to achieve good parallel scalability, and good vector performance of the neutrino transport routines.

Two parallel code versions of PROMETHEUS/VERTEX are currently available. The first uses a two-level hierarchical parallel programming model that exploits instruction level parallelism through vectorization and shared memory parallelism by macrotasking with OpenMP directives. The second code version is similar to the first one, but adds to these two levels also distributed memory parallelism using message passing with MPI.

The nature of the employed algorithms naturally lends itself to a hierarchical programming model because of the fact that directional (operator) splitting is used in both the hydrodynamic as well as the neutrino transport parts of the code. Thus one needs to perform logically independent, lower-dimensional sub-integrations in order to solve a multi-dimensional problem. For instance, the $N_\vartheta \times N_\nu$ and $N_r \times N_\epsilon \times N_\nu$ integrations resulting within the r and ϑ transport sweeps, respectively, can be performed in parallel with coarse granularity. The routines used to perform the lower-dimensional sub-integrations are then completely vectorized.

Figure 4 shows scaling results of the OpenMP code version on an SGI Altix 3700 Bx2 (using Itanium2 CPUs with 6 MB L3 caches). The measurements are for the S and M setups of Table 1. The Thomas solver has been used to invert the Jacobians. The speedup is initially superlinear, while on 64 processors it is close to 60, demonstrating the efficiency of the employed parallelization strategy. Note that static scheduling of the parallel sub-integrations has been applied, because the Altix is a ccNUMA machine which requires a minimization of remote memory references to achieve good scaling. Dynamic scheduling would not guarantee this, although it would actually be preferable from the algorithmic point of view, to obtain optimal load balancing.

Table 1. Some typical setups with different resolutions.

Setup	$N_r^{\rm hyd}$	N_r	N_ϑ	N_ϵ	N_ν
XS	400	234	32	17	3
S	400	234	126	17	3
M	400	234	256	17	3
L	800	468	512	17	3
XL	800	468	512	34	3

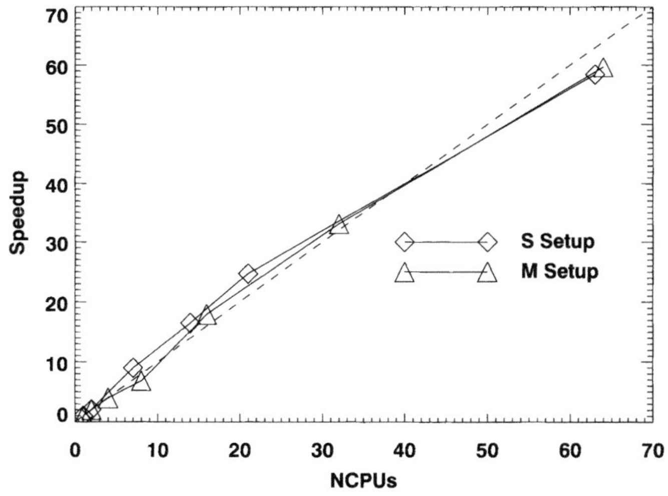

Fig. 4. Scaling of PROMETHEUS/VERTEX on the SGI Altix 3700 Bx2.

Table 2. First measurements of the OpenMP code version on (a single compute node of) an NEC SX-6+ and an NEC SX-8. Times are given in seconds.

Measurements on the SX-6+				
Setup	NCPUs	(avg.) wallclock time/cycle	Speedup	MFLOPs/sec
XS	1	211.25	1.00	2708
XS	4	59.67	3.54	9339
XS	8	34.42	6.14	15844
Measurements on the SX-8				
Setup	NCPUs	(avg.) wallclock time/cycle	Speedup	MFLOPs/sec
XS	1	139.08	1.00	4119
XS	4	39.17	3.43	14181
XS	8	22.75	6.11	23773
S	1	457.71	1.00	4203
S	4	133.43	3.43	13870
S	8	78.43	5.83	22838
M	1	926.14	1.00	4203
M	4	268.29	3.45	13937
M	8	159.14	5.82	22759

The behaviour of the same code on the (cacheless) NEC SX-6+ and NEC SX-8 is shown in Table 2. One can note that (for the same number of processors) the measured speedups are noticeably smaller than on the SGI. Moreover the larger problem setups (with more angular zones) scale *worse* than the smaller ones, indicating that a load imbalance is present. On these "flat memory" machines with their very fast processors a good load balance is apparently much more crucial for obtaining good scalability, and dynamic scheduling of the sub-integrations might have been the better choice. Table 2 also lists the FLOP rates for the entire code (including I/O, initializations and other overhead). The vector performance achieved with the listed setups on a single CPU of the NEC machines is between 26% and 30% of the peak performance. Given that in any case only 17 energy bins have been used in these tests, and that therefore the average vector length achieved in the calculations was only about 110 (on an architecture where vector lengths $\gtrsim 256$ are considered optimal), this computational rate appears quite satisfactory. Improvements are still possible, though, and optimization of the code on NEC machines is in progress.

Acknowledgements

Support from the SFB 375 "Astroparticle Physics" of the Deutsche Forschungsgemeinschaft, and computer time at the HLRS and the Rechenzentrum Garching are acknowledged. We also thank M. Galle and R. Fischer for performing the benchmarks on the NEC machines.

References

1. Rampp, M., Janka, H.T.: Spherically Symmetric Simulation with Boltzmann Neutrino Transport of Core Collapse and Postbounce Evolution of a 15 $M_\odot$ Star. Astrophys. J. **539** (2000) L33–L36
2. Mezzacappa, A., Liebendörfer, M., Messer, O.E., Hix, W.R., Thielemann, F., Bruenn, S.W.: Simulation of the Spherically Symmetric Stellar Core Collapse, Bounce, and Postbounce Evolution of a Star of 13 Solar Masses with Boltzmann Neutrino Transport, and Its Implications for the Supernova Mechanism. Phys. Rev. Letters **86** (2001) 1935–1938
3. Liebendörfer, M., Mezzacappa, A., Thielemann, F., Messer, O.E., Hix, W.R., Bruenn, S.W.: Probing the gravitational well: No supernova explosion in spherical symmetry with general relativistic Boltzmann neutrino transport. Phys. Rev. D **63** (2001) 103004–+
4. Thompson, T.A., Burrows, A., Pinto, P.A.: Shock Breakout in Core-Collapse Supernovae and Its Neutrino Signature. Astrophys. J. **592** (2003) 434–456
5. Bethe, H.A.: Supernova mechanisms. Reviews of Modern Physics **62** (1990) 801–866
6. Burrows, A., Goshy, J.: A Theory of Supernova Explosions. Astrophys. J. **416** (1993) L75
7. Janka, H.T.: Conditions for shock revival by neutrino heating in core-collapse supernovae. Astron. Astrophys. **368** (2001) 527–560
8. Herant, M., Benz, W., Colgate, S.: Postcollapse hydrodynamics of SN 1987A – Two-dimensional simulations of the early evolution. Astrophys. J. **395** (1992) 642–653
9. Herant, M., Benz, W., Hix, W.R., Fryer, C.L., Colgate, S.A.: Inside the supernova: A powerful convective engine. Astrophys. J. **435** (1994) 339
10. Burrows, A., Hayes, J., Fryxell, B.A.: On the nature of core-collapse supernova explosions. Astrophys. J. **450** (1995) 830
11. Janka, H.T., Müller, E.: Neutrino heating, convection, and the mechanism of Type-II supernova explosions. Astron. Astrophys. **306** (1996) 167–+
12. Thompson, C.: Accretional Heating of Asymmetric Supernova Cores. Astrophys. J. **534** (2000) 915–933
13. Foglizzo, T.: Non-radial instabilities of isothermal Bondi accretion with a shock: Vortical-acoustic cycle vs. post-shock acceleration. Astron. Astrophys. **392** (2002) 353–368
14. Blondin, J.M., Mezzacappa, A., DeMarino, C.: Stability of Standing Accretion Shocks, with an Eye toward Core-Collapse Supernovae. Astrophys. J. **584** (2003) 971–980
15. Scheck, L., Plewa, T., Janka, H.T., Kifonidis, K., Müller, E.: Pulsar Recoil by Large-Scale Anisotropies in Supernova Explosions. Phys. Rev. Letters **92** (2004) 011103–+
16. Keil, W., Janka, H.T., Mueller, E.: Ledoux Convection in Protoneutron Stars – A Clue to Supernova Nucleosynthesis? Astrophys. J. **473** (1996) L111
17. Burrows, A., Lattimer, J.M.: The birth of neutron stars. Astrophys. J. **307** (1986) 178–196

18. Pons, J.A., Reddy, S., Prakash, M., Lattimer, J.M., Miralles, J.A.: Evolution of Proto-Neutron Stars. Astrophys. J. **513** (1999) 780–804
19. Janka, H.T., Mönchmeyer, R.: Anisotropic neutrino emission from rotating protoneutron stars. Astron. Astrophys. **209** (1989) L5–L8
20. Janka, H.T., Mönchmeyer, R.: Hydrostatic post bounce configurations of collapsed rotating iron cores – Neutrino emission. Astron. Astrophys. **226** (1989) 69–87
21. Shimizu, T.M., Ebisuzaki, T., Sato, K., Yamada, S.: Effect of Anisotropic Neutrino Radiation on Supernova Explosion Energy. Astrophys. J. **552** (2001) 756–781
22. Kotake, K., Yamada, S., Sato, K.: Anisotropic Neutrino Radiation in Rotational Core Collapse. Astrophys. J. **595** (2003) 304–316
23. Fryer, C.L.: Mass Limits For Black Hole Formation. Astrophys. J. **522** (1999) 413–418
24. Fryer, C.L., Heger, A.: Core-Collapse Simulations of Rotating Stars. Astrophys. J. **541** (2000) 1033–1050
25. Fryer, C.L., Warren, M.S.: Modeling Core-Collapse Supernovae in Three Dimensions. Astrophys. J. **574** (2002) L65–L68
26. Fryer, C.L., Warren, M.S.: The Collapse of Rotating Massive Stars in Three Dimensions. Astrophys. J. **601** (2004) 391–404
27. Rampp, M., Janka, H.T.: Radiation hydrodynamics with neutrinos. Variable Eddington factor method for core-collapse supernova simulations. Astron. Astrophys. **396** (2002) 361–392
28. Buras, R., Rampp, M., Janka, H.T., Kifonidis, K.: Improved Models of Stellar Core Collapse and Still No Explosions: What Is Missing? Phys. Rev. Letters **90** (2003) 241101–+
29. Janka, H.T., Buras, R., Kifonidis, K., Marek, A., Rampp, M.: Core-Collapse Supernovae at the Threshold. In Marcaide, J.M., Weiler, K.W., eds.: Supernovae, Procs. of the IAU Coll. 192, Berlin, Springer (2004)
30. Buras, R., Rampp, M., Janka, H.T., Kifonidis, K., Takahashi, K., Horowitz, C.J.: Two-dimensional hydrodynamic core collapse supernova simulations with spectral neutrino transport. Astron. Astrophys. (2006), to appear
31. Müller, E., Rampp, M., Buras, R., Janka, H.T., Shoemaker, D.H.: Toward Gravitational Wave Signals from Realistic Core-Collapse Supernova Models. Astrophys. J. **603** (2004) 221–230
32. Lattimer, J.M., Swesty, F.D.: A generalized equation of state for hot, dense manner. Nuclear Physics A **535** (1991) 331–+
33. Shen, H., Toki, H., Oyamatsu, K., Sumiyoshi, K.: Relativistic Equation of State of Nuclear Matter for Supernova Explosion. Progress of Theoretical Physics **100** (1998) 1013–1031
34. Hillebrandt, W., Wolff, R.G.: Models of Type II Supernova Explosions. In Arnett, W.D., Truran, J.W., eds.: Nucleosynthesis: Challenges and New Developments, Chicago, University of Chicago Press (1985) 131
35. Marek, A.: The effects of the nuclear equation of state on stellar core collapse and supernova evolution. Diplomarbeit, Technische Universität München (2003)
36. Mihalas, D., Weibel Mihalas, B.: Foundations of Radiation Hydrodynamics. Oxford University Press, Oxford (1984)

Applications IV: Turbulence Simulation

Statistics and Intermittency of Developed Channel Flows: a Grand Challenge in Turbulence Modeling and Simulation

Kamen N. Beronov[1], Franz Durst[1], Nagihan Özyilmaz[1], and Peter Lammers[2]

[1] Institute of Fluid Mechanics (LSTM), University Erlangen-Nürnberg, Cauerstraße 4, D-91058 Erlangen, Germany, {kberonov,durst,noezyilm}@lstm.uni-erlangen.de,
[2] High Performance Computing Center Stuttgart (HLRS), Nobelstraße 19, D-70569 Stuttgart, Germany

Abstract Studying and modeling turbulence in wall-bounded flows is important in many engineering fields, such as transportation, power generation or chemical engineering. Despite its long history, it remains disputable even in its basic aspects and even if only simple flow types are considered. Focusing on the best studied flow type, which has also direct applications, we argue that not only its theoretical description, but also its experimental measurement and numerical simulation are objectively limited in range and precision, and that it is necessary to bridge gaps between parameter ranges that are covered by different approaches. Currently, this can only be achieved by expanding the range of numerical simulations, a grand challenge even for the most powerful computational resources just becoming available. The required setup and desired output of such simulations are specified, along with estimates of the computing effort on the NEC SX-8 supercomputer at HLRS.

1 Introduction

Among the millennium year events, one important for mathematical physics was the formulation of several "grand challenge" problems, which remain unsolved after many decades of efforts and are crucial for building a stable knowledge basis. One of these problems concerns the existence of solutions to the three-dimensional Navier-Stokes equations. The fundamental understanding of the different aspects of turbulent dynamics generated by these equations, however, is a much more difficult problem, remaining unsolved after more than 100 years of great effort and ever growing range of applications in engineering and the natural sciences. Of great practical relevance is the understanding of turbulence generation and regeneration in the vicinity of solid boundaries: smooth, rough, or patterned. Starting from climate research and weather prediction, covering aeronautics and automotive engineering, chemical and machine engineering, and nowadays penetrating into high mass-flow microfluidics, the issues of near-wall turbulence are omnipresent in the research and design practice. But still, as shown by some examples below, an adequate understanding is lacking and the computational practice is fraught with controversy, misunderstanding and misuse of approximations.

It was only during the last 20 years that a detailed qualitative understanding of the near-wall turbulent dynamics could be established. With the advent of reliable methods for direct numerical simulation (DNS) and the continuous growth of computing power, the DNS of some low- to moderate-Reynolds-number wall-bounded turbulent flows became possible and provided detailed quantitative knowledge of the turbulent flow layers which are closest to the wall and are the earliest to approach their asymptotic state with respect to growing Reynolds number $Re \to \infty$. In the last few years, most basic questions concerning the nonlinear, self-sustaining features of turbulence in the viscous sublayer [19] and in the buffer layer adjacent to it, including the "wall cycle" [10], were resolved by a combination of analytics and data analysis of DNS results. Following on the agenda are now the nature and characteristics of the next adjacent layer [20], in which no fixed characteristic scale appears to exist and the relevant length scales, namely the distance to the wall and the viscous length, are spatially varying and disparate from each other. This is reminiscent of the conditions for the existence of an inertial range in homogeneous turbulence, but the presence of shear and inhomogeneity complicate the issues. This layer is usually modeled as having a logarithmic mean velocity profile, but this is not sufficient to characterize the turbulence, is not valid at lower, still turbulent Reynolds numbers, and is still vehemently disputed in view of the very competitive performance of power-law rather than logarithmic laws. Both types reflect in a way the self-similarity of turbulent flow structures, which had been long hypothesized and has now been documented in the literature, see [20] for references.

This "logarithmic layer" is passive with respect to the turbulence sustaining "wall cycle" [10, 20], somewhat like the "inertial range" with respect to the "energy containing range" in homogeneous turbulence. Precisely because of this and its related self-similarity features, it should be easier to model. In practice, this has long be used in the "wall function" approach of treating near-wall turbulence in numerics by assuming log-law mean velocity. Its counterpart in homogeneous turbulence are the inertial-range cut-off models underlying all subgrid-scale modeling (SGSM) for large-eddy simulation (LES) methods currently in use. Both SGSM and wall-function models can be related to corresponding eddy-viscosity models. In the wall-bounded case, however, the effect of distance to the wall and, through it, of Reynolds number, is important and no ultimate quantitative models are available. This is due mostly to the still very great difficulty in simulating wall-bounded flows at sufficiently high Reynolds numbers – the first reports of DNS in this range [20] estimate this as one of the great computational challenges that will be addressed in the years 2005–2010.

Some of the theoretical and practical modeling issues that will be clarified in this international and competitive effort, including the ones mentioned already, are presented in quantitative detail in Sect. 2. Some critical numerical aspects of these grand challenge DNS projects are presented in Sect. 3, leading to the suggestion of lattice Boltzmann methods as the methods of choice for such very large scale simulations and to estimates that show such a DNS project to be practicable on the NEC SX-8 at HLRS.

2 State of Knowledge

At the most fundamental level, the physical issue of interest here is the interplay of two length scales, the intrinsic dissipative scale and the distance to the wall, when these are sufficiently different from each other, as well as two time scales, the mean flow shear rate, which generally enhances turbulence, and the rate of turbulent energy dissipation. The maximum mean flow shear occurs at the wall; it is customary to use the corresponding strain rate to define a "friction velocity" u_τ and, over the Newtonian viscosity, a viscous length δ_ν. These "wall units" are used to nondimensionalize all hydrodynamic quantities in the "inner scaling," such as $\nu^+ = 1$, velocity $u^+ = u/u_\tau$, and "friction Reynolds number"

$$Re_\tau = H/\delta_\nu = Hu_\tau/\nu \tag{1}$$

where H is a cross-channel length scale, defined as the radius for circular pipes and half the channel width for flows between two parallel planes. In the alternative "outer scaling," lengths are measured in units of H and velocities in units of $\bar{U}$, the mean velocity over the full cross-section of the channel, or of U_c, the "centerline" velocity (in channels) or "free stream" velocity (in boundary layers). The corresponding Reynolds numbers are

$$Re_m = 2H\bar{U}/\nu\,, \qquad Re_c = HU_c/\nu \tag{2}$$

The interplay of the "inner" and "outer" dynamics is reflected by the friction factor, the bulk quantity of prime engineering interest:

$$c_f(Re) = 2\left(u_\tau/\bar{U}\right)^2\,, \qquad C_f(Re) = 2(u_\tau/U_c)^2\,, \tag{3}$$

There are two competing approximations for C_f, both obtained from data on developed turbulence in straight circular pipes only, respectively due to Blasius (1905) and Prandtl (1930):

$$C_f(Re) = C_B Re_m^{-\beta}\,, \qquad \beta = 1/4\,, \quad C_B \approx 0.3168/4 \tag{4}$$

$$C_f^{1/2}\left(\log_{10}\left(C_f^{1/2} Re_m\right) - C_L\right) = 1/C_P\,, \quad C_P \approx 4\,, \quad C_L \approx 0.4\,, \tag{5}$$

The value for C_B given above is taken from the extensive data survey [4], including channels of various cross-sectional shape.

The Blasius formula is precise for Re_m below 10^5, while the Prandtl formula is precise for $Re_m > 10^4$. Both match the data well in the overlap of their range

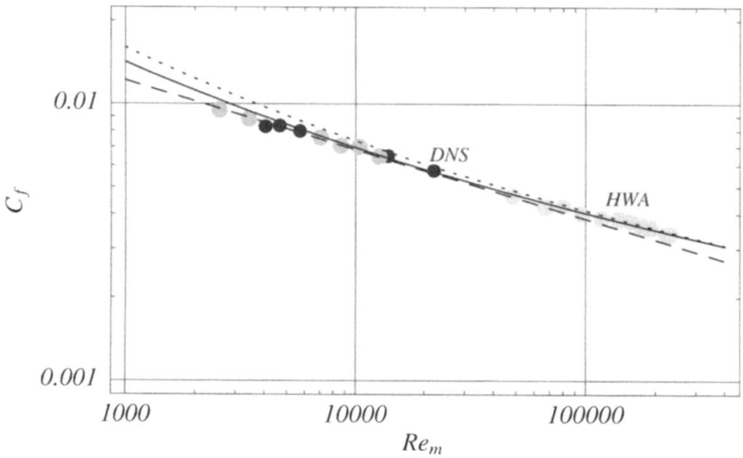

Fig. 1. Friction factor data for plane channel turbulence. Dotted line: Blasius formula (4) for pipe flow. in coeff. Dashed line: Blasius formula modified for plane channel flow, with $C_B = 0.0685$. Solid line: Prandtl formula modified for plane channel flow, whose $C_P = 4.25$ differs from (5). The low–Re data from DNS (black points) and laser-Doppler anemometry (dark gray [13, 12]) show that the former type of formula is adequate for $Re_m < 2.10^4$. The high–Re data from hot-wire anemometry (light gray [14]) support the latter type of formula for $Re_m > 5.10^4$

of validity. A similar situation, with only slightly different numerical coefficients, can be observed for developed plane channel turbulence, as well, as illustrated in Fig. 1. While the abundance of measurements for pipe flows and zero pressure gradient boundary layers allows to cover the overlap range with data points and to reliably extract the approximation coefficients, the available data on plane channel flows remains insufficient. As seen in Fig. 1, data are lacking precisely in the most interesting parameter range, when both types of approximations match each other. There are no simulation data available yet, which could confirm the Prandtl-type approximation for plane channel flows, even not over the overlap range where it matches a Blasius-type formula.

2.1 Mean Flow: an Ongoing Controversy

The two different scalings of the friction factor reflect the different scaling, with growing distance from the wall, of the mean velocity profile at different Reynolds numbers. It is a standard statement found in textbooks [3] that there is, at sufficiently large Re, a "self-similarity layer" as described above and situated between the near-wall region (consisting of the viscous and buffer layers, approximately located at $y^+ < 10$ and $10 < y^+ < 70$, respectively) and the core flow (whose description and location depends significantly on the flow geometry), and that this layer is characterized by a logarithmic mean velocity profile:

$$\bar{U}^+(y^+, Re) = \ln(y^+) / \kappa + B \tag{6}$$

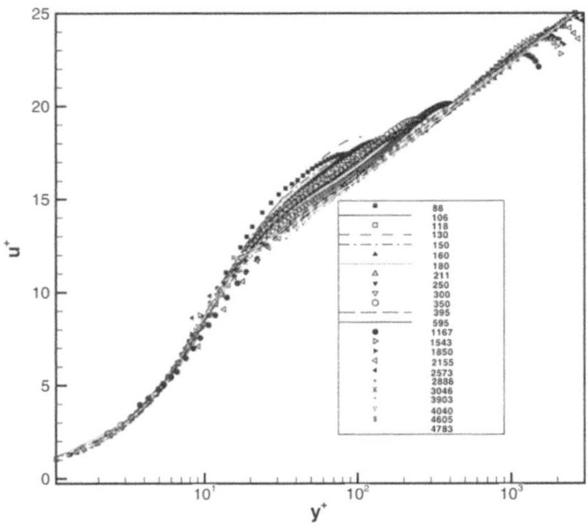

Fig. 2. Top: mean flow profiles at different Reynolds numbers. Bottom: diagnostic functions for power-law and log-law (8)

It appears intuitive that at lower Re the same layer is smaller in both inner and outer variables, but it is not so well known except in the specialized research literature [20], that for $Re_\tau < 1000$ no logarithmic layer exists, while turbulence at a smooth wall is self-sustained already at $Re_\tau \approx 100$. In fact, a log-layer is generally assumed in standard engineering estimates and even used in "low-Reynolds-number" turbulence models in commercial CFD software! The power-law Blasius formula suggests, however, that the mean velocity profile $\bar{U}^+(y^+, Re)$ at low Re_τ is for the most part close to a power law:

$$\bar{U}^+(y^+, Re) = (y^+)^{\gamma(Re)} A(Re) \tag{7}$$

whereby $\beta = 1/4$ in the Blasius formula corresponds to $\gamma = 1/7 \approx 0.143$. It was soon recognized on the basis of detailed measurements [1] that at least one of the parameters in (7) must be allowed to have a Reynolds number dependence. A data fitting based on adjusting $\gamma(Re)$ was already described in [1].

Recently, the general form of (7) has been reintroduced [5], based on general theoretical reasoning and on reprocessing circular pipe turbulence data, first from the original source [1] and later from modern measurements [6]. It is not only claimed that both parameters are simple algebraic functions of $\ln(Re_\tau)$, but also the particular functional forms and the corresponding empiric coefficients are fixed [5]. Moreover, the same functional form is shown to provide good fits also to a variety of zero pressure gradient boundary layer data with rather disparate Reynolds numbers and quality of the free stream turbulence [9]. The overall claim in these works is that the power-law form (7) is universally valid, even at very large Reynolds numbers and for all kinds of canonical wall-bounded turbulence, and that no finite limit corresponding to κ in (6) exists with $Re \to \infty$, as required in the derivation of the log-law. Thus, the log-law is completely rejected and replaced by (7) with particular forms for $\gamma(Re)$ and $A(Re)$ depending, at least quantitatively, also on the flow geometry and the free stream turbulence characteristics. An interpretation of the previously observed log-law as envelope to families of velocity profile curves is given; no comment is provided on the success of the Prandtl formula for C_f, which is based on the assumption of a logarithmic profile over most of the channel or boundary layer width.

The non-universal picture emerging from these works is intellectually less satisfactory but not necessarily less compatible with observations on the dependence of turbulence statistics on far-field influences and Reynolds-number effects. The controversy about which kind of law is the correct one is still going on and careful statistical analyses of available data have not been able to discriminate between the two on the basis of error minimization. It has been noted, however, that at least two different power laws are required in general, in order to cover the most of the channel cross-section width, an observation already present in [9].

We have advanced a more pragmatic view: The coexistence of the Blasius and Prandtl type of formulas, justified by abundant data, as well as direct observations of mean profiles, suggest that a log-law is present only at Re_τ above some threshold, which for plane channel flows lies between 1200 and 1800, approximately corresponding to the overlap region between the mentioned two types of C_f formulas. It is recalled that both the logarithmic and the power-law scaling of the mean velocity with wall distance can be rigorously derived [8] and thus may well coexist in one profile over different parts of a cross-section.

It is furthermore assumed, in concord with standard theory [3], that a high-Re limit of the profile $\bar{U}(y^+, Re_\tau)$ exists for any fixed y^+. These two assumptions suggest the existence of a power-law portion of the $\bar{U}$ profile at lower y^+ and an adjacent logarithmic type of profile at higher y^+. The latter can of course be present only for sufficiently large Re_τ. The simultaneous presence of these two, smoothly joined positions of the mean velocity profile was verified on the basis of a collection of experimental and DNS data of various origin and covering a wide range of Reynolds numbers. This is illustrated in Fig. 2 using the diagnostic functions, Γ and Ξ, which are constant in y^+ regions where the mean velocity profile is given by a power-law and a log-law, respectively:

$$\Gamma(y^+, Re) = \frac{y^+}{\bar{U}^+}\frac{d\bar{U}^+}{dy^+}\,, \qquad \Xi(y^+, Re) = y^+\,\frac{d\bar{U}^+}{dy^+}\,. \tag{8}$$

The large-Re universality of the profile is assured by the existence of a finite limit for the power-law portion, contrary to the statements in [5, 9]. It was found that the parametric dependence on Re is indeed a simple algebraic function in $\ln(Re_\tau)$, as suggested in [5], but that it suffices to have only one of the parameters vary. A very good fit is nevertheless possible, since only a fixed y^+ range is being fitted and no attempt is made to cover a range growing with Re_τ as in [1, 5, 9]. And contrary to [1], the power-law exponent is not allowed to vary but is instead estimated by minimizing the statistical error of available data. The result is illustrated in Fig. 3:

$$\gamma \approx 0.154\,, \qquad A(Re) \approx 8 + 500/\ln(Re_\tau)^4 \tag{9}$$

By a similar procedure, it was estimated that the power-law range in y^+ extends approximately between 70 and 150, then smoothly connecting over the range 150-250 to a pure log-law range in y^+. To describe reliably this transition, very

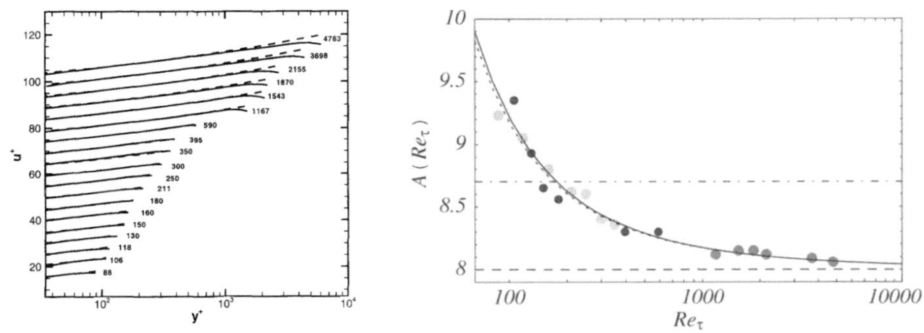

Fig. 3. Fitting a power law in the adjacent layer, $70 < y^+ < 150$, next to the buffer layer. Left: approximations to individual mean profiles at separate Reynolds numbers, using (7) with fixed γ as given by (9) and $A(Re)$ as the only fitting parameter. Right: the dependence $A(Re)$ thus obtained is compared to the formula given in (9)

reliable data are required. Unfortunately, hot-wire anemometry (HWA) data are not very precise and laser-Doppler anemometry (LDA) measurements are difficult to obtain at high Re in the mentioned y^+ range, which is then technically rather close to the wall. DNS could provide very reliable data, but are very expensive and, for that reason, still practically unavailable. There is only one high-quality simulation known [20], which approaches with $Re_\tau \approx 950$ the transition Reynolds number, but it is still too low to feature a true log-layer of even minimal extent. To enter the asymptotic regime with developed log-layer and almost converged power-law region, it is necessary to simulate reliably $Re_\tau \approx 2000$ and higher. What reliability of channel turbulence DNS implies is discussed in Sect. 3.

2.2 Eddy Viscosity

The eddy viscosity ν_T is a an indispensable attribute of all modern RANS and LES models of practical relevance for CFD. Its definition is usually based on the dissipation rate ε of turbulent kinetic energy k, assuming complete analogy with the dissipation in laminar flows:

$$\varepsilon = \nu_T \left(\nabla \bar{U} + (\nabla \bar{U})^\dagger\right) : (\nabla \bar{U}) \tag{10}$$

$$\nu_T{}^+ = \nu_T/\nu = \frac{dy^+}{d\bar{U}^+}\left(1 - \frac{y^+}{Re_\tau}\right) - 1 \tag{11}$$

The latter direct relation between eddy viscosity and mean velocity gradient is an exact consequence of the mean momentum balance equation. It is customary, however, to nondimensionalize the eddy viscosity, anticipating simple scaling behaviour:

$$\nu_T \varepsilon / k^2 = \nu_T{}^+ \varepsilon^+ \left(k^+\right)^{-2} = C_\nu(y^+, Re_\tau) \tag{12}$$

$$70 < y^+ \quad : \; C_\nu \approx 0.09 \tag{13}$$

$$0.4 < y/H \; : \; C_\nu \approx 0.09 \left(k^+\right)^{-2/5} Re_\tau \tag{14}$$

The former approximation (13) is a standard in RANS computations of turbulent flows, with various prescriptions for the "constant" C_ν on the right, mostly in the range 0.086–0.10 for channel flows. The latter approximation (14) is a proposal we have put forward on the basis of data analysis from several moderate–Re_τ DNS of channel flow turbulence. Evidence for both approximations, based on the same data bases but considering different layers of the channel simulated flows, is presented in Fig. 4. It is clear that substantially higher Re than the maximal one shown in that Figure, $Re_\tau \approx 640$, are required in order to quantify the high-Re asymptote and, more importantly, any possible qualitative effect of transition to the log-law regime, which may influence the above C_ν scalings, especially (14). The latter issue requires at least $Re_\tau \approx 2000$, simulated in computational domains many times longer in outer units H than the DNS reported in [20] and its corresponding references. Evidence from HWA measurements [14] indicates that the power-law in (14) persists in the log-law range, allowing to hope that it can be verified with a very limited number of high-Re DNS, e.g. adding only one at $Re_\tau \approx 4000$.

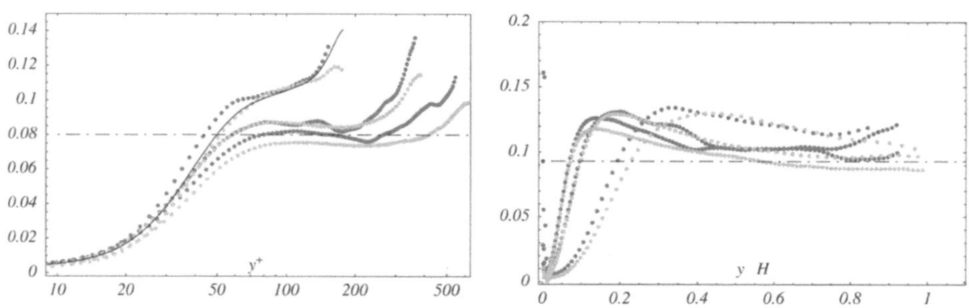

Fig. 4. Normalized eddy viscosity C_ν defined in (12), from DNS data by different methods: LBM [18, 16], Chebyshev pseudospectral [7], finite difference [11]. Left: inner scaling, dashed line indicates 0.08 instead of the r.h.s. of (13). Right: outer scaling, dashed line indicates 0.093 instead of the constant in (14)

2.3 Fluctuating Velocities

An issue of significant engineering interest beside the mean flow properties discussed so far, including the eddy viscosity, is the characterization of Reynolds stresses. The same level of precision as for the mean velocity is requested, i.e. (y^+, Re_τ) and $(y/H, Re_m)$ profiles. It is known from the experimental literature that even at higher Re_τ than accessible to DNS so far, a significant Re-dependence remains, especially in the intensity of the streamwise fluctuating velocity component. At the same time, it is known [2, 20] that a noticeable contribution to these amplitudes comes from "passive" flow structures, especially far from the wall. The separation of such structures from the "intrinsic" wall cycle structures is only practicable by filtering DNS data [10]. It is also necessary to assure sufficiently long computational domains to eliminate the "self-excitation" of these fluctuations, see Sect. 3.2 and [20].

Higher-order moments of individual velocity components, starting with the flatness and kurtosis, are of continuing interest to physicists. Although the deviation from Gaussianity diminishes farther from the wall, it is qualitatively present throughout the channel and influences the turbulence structure. The most general approach to describing such statistics is the estimation of probability density functions (PDFs) for the individual velocity components as well as joined probability densities, from which the moments of individual velocity components and cross-correlations can be determined, respectively. In principle, all convergent moments can be determined from the PDF, but the latter need to be known with increasing precision over value ranges of increasing width as the order of the moment increases. In practice, it is difficult to obtain statistically converged estimates of high-order moments or, equivalently, of far tails of PDFs, from DNS data. The quality of statistics grows with increasing number of simulation grid points and time steps. Thus, improved higher-order statistics can be obtained from DNS at higher *Re* as a byproduct of the necessarily increased computational grid size and simulation time.

This effect is illustrated in Fig. 5, where the LBM at $Re_\tau = 180$ [18] with several times more grid points than the corresponding pseudospectral simulations at $178 < Re_\tau < 588$ [7] captures the tails of the wall/normal velocity PDF $p(v)$ up to significantly larger values. The comparison serves also as verification of the LBM code and as an indication that normalized lower-oder moments converge fast with growing *Re*. It is not clear, however, whether the PDFs in the power-law

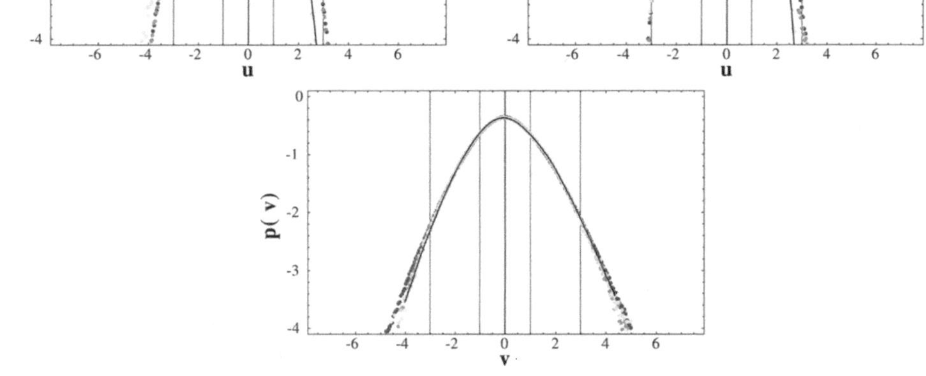

Fig. 5. Estimates of probability density functions for fluctuating velocity components, from DNS with LBM and spectral methods. Comparison of LBM (lines [18]) and Chebyshev pseudospectral (points [7]) DNS of plane channel turbulence at Re_τ between 130 and 590

and log-law range scale with inner variables, as is the case at the close distances to the wall shown in Fig. 5, or with outer variables, or most probably, present a mixture of both. To clarify this issue, especially if the latter case of mixed scaling holds, a number of reliable higher-*Re* DNS are required.

2.4 Streamwise Correlations

Beyond the single-point velocity statistics discussed so far, it is important to investigate also multi-point, multi-time correlations in order to clarify the turbulence structure. Already the simplest example of two-point, single-time auto- and cross-correlations (or their corresponding Fourier spectra, see [20] and its corresponding references) reveals the existence of very long streamwise correlations. Corresponding very long flow structures have been documented in experiments and numerical simulations during the last 10 years. Their origin and influence on the turbulent balances have not been entirely clarified. It was recognized, however, that their influence is substantial, in particular on streamwise velocity statistics, and that domains of very large streamwise extent, at least $L_x \approx 25H$, have to be considered both in experimental and in DNS set-up in order to minimize domain size artifacts.

From experimental data [17] and own DNS [18] we have found that the first zero-crossing of the autocorrelation function of the streamwise fluctuating velocity component scales as $y^{2/3}$ and reaches maxima at the channel centerline, which grow in absolute value with growing Reynolds number, cf. Fig. 6. At $Re_\tau = 180$ it is about $26H$, but at $Re_\tau \approx 2200$ it is already about $36H$, as seen in Fig. 6. These are values characteristic of the longest turbulent flow structures found at the respective Reynolds numbers. Analysis of the shape of autocorrelation functions shows that, indeed, these are relatively weak, "passive structures" in the sense of Townsend [2]. It is important to quantify the mentioned *Re*-dependence of maximal structure length. To that end, additional DNS and experimental data are required, at least over $1000 < Re_\tau < 4000$.

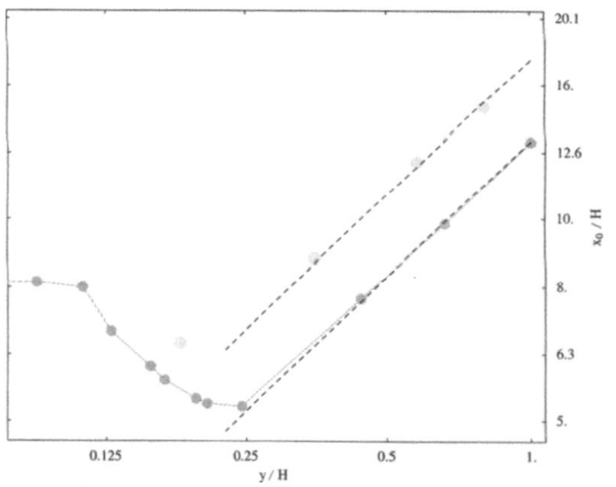

Fig. 6. zero-crossing: dashed lines show $\sim y^{2/3}$ scaling

3 Direct Numerical Simulation

The preceding discussion of the state of knowledge about the turbulence structure in wall-bounded flows focused on the layer adjacent to the buffer layer and usually referred to as the log-layer. It was shown that using a simple log-law there is incorrect not only when the Reynolds number is relatively law, $Re_\tau < 1000$, but also at high Re when the layer can be decomposed into a power-law layer immediately next to the buffer layer, and a smoothly attached true log-law layer farther from the wall. A clear demonstration of this kind of layer structure is still pending, since no reliable DNS over a sufficiently long computational domain at $Re_\tau > 1000$ is available so far.

3.1 The Physical Model: Plane Channel Flow

At such high Re_τ, the similarity between plane channel and circular pipe flow can be expected to be very close, at least over the near-wall and the power-law layers, i.e. the spatial ranges of prime interest here. The standard analytical and DNS model used to investigate fully developed turbulence is a periodic domain in the streamwise direction. What corresponding spatial period length would be sufficient to avoid self-excitations is discussed in Sect. 2.4 and 3.2. The flow is driven by a prescribed, constant in time, streamwise pressure gradient or mass flow rate. Incompressible flow with constant density and constant Newtonian viscosity is assumed.

To maximize Re_τ and minimize geometrical dependencies in the near-wall layers (the increase in Re contributes to the reduction of such dependencies) within the framework of the above specifications, it is computationally advantageous to simulate plane channel flow. Periodicity is thus assumed also in the spanwise direction, perpendicular to the wall-normal and to the streamwise directions. For ease of implementation and of organizing the initial transient in the simulation, a constant pressure gradient forcing is chosen.

The Reynolds number defined in (1) should, according to the analysis of open problems in Sect. 2, be chosen at several values in the range $800 \leq Re_\tau \leq 4000$ in order to cover the transition to the log-law range and at least two cases clearly in that range. A possible Re_τ sequence with only four members is thus e.g. 800–1000, 1200–1500, 1800–2000, and 3600–4000. The quantities of interest are those listed in Sect. 2, including all Reynolds stresses, the dissipation rate ε, and the two-point velocity correlations. Also of interest are the vorticity component statistics paralleling the mentioned velocity statistics, as well as joint PDFs of velocity, vorticity and strain components and of pressure and pressure gradient.

The characterization of self-similarity and of correlations scaling with distance form the wall in inner and outer units is of currently prime scientific interest. It is e.g. of practical interest for CFD modeling to know if a data collapse for an extended version of Fig. 6 can be achieved in inner variables. A quantification of the degree of residual Reynolds-number dependence in velocity and vorticity momenta, dissipation and other energy balance terms (cf. Sect. 2.2), would provide a new, decisive impetus to turbulence modeling.

3.2 Required Computational Domain

In a detailed parametric study with LBM simulations in domains of increasing streamwise sizes, the present authors [18] have found that even $L_x > 30H$ is required. The size $25H$ cited above, used in [2, 20] and some of the simulations cited therein, was found to be insufficient over parts of the channel cross-section, see Fig. 7. The streamwise length of $L_x = 32H$ used in [18] appears acceptable for the range $Re_\tau \leq 1000$ simulated so far [20], but as far as the pipe flow data of Fig. 6 can be carried over to plane channels, it will no longer suffice for $Re_\tau \geq 2000$. To cover the Re-range suggested above, it is proposed to use $L_x = 36H$ for $1500 \leq Re_\tau \leq 2000$ and then $L_x = 40H$ up to $Re_\tau \leq 4000$.

The spanwise spatial period length L_z is taken in the channel DNS literature within $3 \leq L_z/H \leq 8$ to assure unaffected spanwise correlations. But the main statistics include averaging over the spanwise direction and do not include directly the spanwise velocity component. The present authors' experience [15, 16] is that these statistics are not strongly influenced even if only $L_z/H = 2$ (square cross-section of the computational box) is chosen. The savings are to be used to assure a sufficiently long streamwise box size.

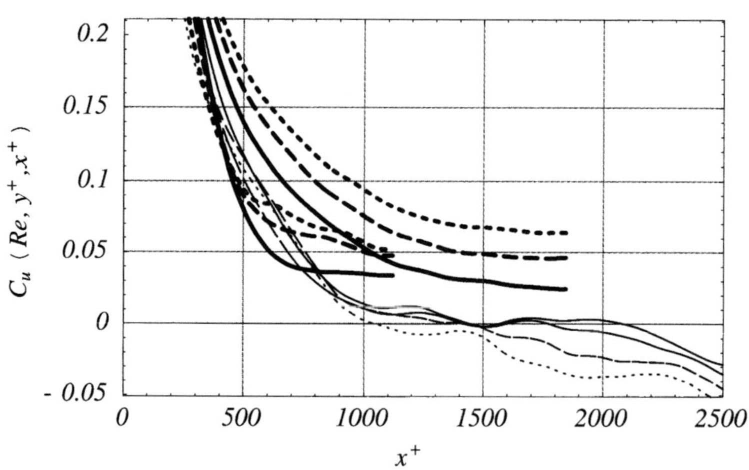

Fig. 7. The long-distance tails of the auto-correlation function of the streamwise fluctuating velocity component at different locations off the wall in plane channel DNS at $Re_\tau = 180$ with LBM [17] and pseudospectral [7] and at $Re_\tau = 590$ [7]. The LBM computational domain had considerably longer streamwise dimension in channel height units, providing the longer tails of computed correlations seen here. Near the wall (left) the short-range correlations obtained by both methods at $Re_\tau = 180$ agree but the long-range data of [7] show spurious "infinite correlation" (see [20] for a comment). In the core flow (right) the spurious correlations are still present in the $Re_\tau = 590$ simulation data base of [7] but no longer in their $Re_\tau = 180$ data base [7]. The latter are still plagued, however, this time by oscillations in their short-range part – compare to the LBM data for the same wall distances

3.3 Method of Choice: Lattice Boltzmann

Computations can be parallelized by optimal domain decomposition and the application of a computational method optimal for large grids and large number of subdomains. The very long computational domain is easily split in the streamwise direction into 16 or 32 equally sized subdomains. This is already a sufficient granularity for distribution among the computing nodes of the NEC SX-8 at HLRS. A numerical method with a theoretically minimal, 2D linear cost of communication per time step, is the standard family of lattice Boltzmann methods (LBM). A related advantage of LBM in computing nominally incompressible flows is that its algorithm involves no Poisson solve steps, since the method is fully explicit (which is known to pose no essential restriction in DNS of intense turbulence) and since it attains (only approximately) incompressibility dynamically, as in the long known artificial compressibility methods. This implies that nonlocal operations with their costly communication are not required for time marching, only when statistics for two-point correlations need to be accumulated.

The LBM code BEST was developed, validated and extensively tested for plane channel and related turbulent flows by the present authors [15, 16, 18]. It was theoretically estimated that at sufficiently large block sizes LBM will have a performance advantage over the standard pseudospectral methods for plane channel DNS using the same grid size. It was then shown by using BEST that the cross-over takes place already at blocks of 128^3 grid points. It was further estimated on the basis of Kolmogorov length data in the vicinity of the wall and on the usual stability criterion from homogeneous turbulence DNS, that the uniform step of the LBM grid at the wall should not exceed 2.4 wall units. The practical limit found with BEST was about 2.3 units.

Thus a uniform grid will need a cross-section of about 700 points for a DNS at $Re_\tau = 800$ and 480 points for $Re_\tau = 580$ (close to the highest Re in [7]). In the latter case, the computational grid will have 7680 points in the streamwise direction. Running such a grid on 8 nodes of the SX-8 was found to require less than 90 minutes for 10^4 iterations, which correspond approximately to the time scale $H/\bar{U}$. One flow-through time $t_F = L_x/\bar{U}$ for the whole very long box will typically require 30 times that many iterations. Following [20] and own experience, at least $10\,t_F$ should be allowed to compute statistics and 2–3 times that much for the initial transient. Thus, the whole simulation can be completed within one week on 8 dedicated nodes.

References

1. Nikuradse, J.: Gesetzmäßigkeiten der turbulenten Strömung in glatten Rohren. Forschungsheft **359**, Kaiser–Wilhelm–Institut für Strömungsforschung, Göttingen (1932)
2. Townsend, A.A.: The structure of turbulent shear flow. Cambridge University Press (1976)

3. Schlichting, H., Gersten, K.: Grenzschicht–Theorie, 9. bearbeitete und erweiterte Ausgabe, Springer, Berlin (1997)
4. Dean, R.B.: Reynolds number dependence of skin friction and other bulk flow variables in two-dimensional rectangular duct flow. *J. Fluids Eng. Trans. ASME* **100**, 215–223 (1978)
5. Barenblatt, G.I., Chorin, A.J., Prostokishin, V.M.: Scaling laws for fully developed turbulent flow in pipes: Discussion of experimental data. *Proc. Natl. Acad. Sci. USA* **94**, 773–776 (1997)
6. Zagarola, M.V., Smits, A.J.: Scaling of the mean velocity profile for turbulent pipe flow. *Phys. Rev. Lett.* **78**(2), 239–242 (1997)
7. Moser, R., Kim, J., Mansour, N.N.: Direct numerical simulation of turbulent channel flow up to $Re_\tau = 590$. *Phys. Fluids* **11**, 943–945 (1999)
8. Oberlack, M.: Similarity in non-rotating and rotating turbulent pipe flows. *J. Fluid Mech.* **379**, 1–22 (1999)
9. Barenblatt, G.I., Chorin, A.J., Prostokishin, V.M.: Self-similar intermediate structures in turbulent boundary layers at large Reynolds numbers. *J. Fluid Mech.* **410**, 263–283 (2000)
10. Jiménez, J., Simens, M.P.: Low-dimensional dynamics of a turbulent wall flow. *J. Fluid Mech.* **435**, 81–91 (2001)
11. Abe, H., Kawamura, H., Matsuo, Y.: Direct numerical simulation of a fully developed turbulent channel flow with respect to the Reynolds number dependence. *Trans. ASME J. Fluids Eng.* **123**, 382–393 (2001)
12. Fischer, M., Jovanović, J., Durst, F.: Reynolds number effects in the near–wall region of turbulent channel flows. *Phys. Fluids* **13**(6), 1755–1767 (2001)
13. Fischer, M.: Turbulente wandgebundene Strömungen bei kleinen Reynoldszahlen. Ph.D. Thesis, University Erlangen-Nürnberg (1999)
14. Zanoun, E.-S. M.: Answers to some open questions in wall-bounded laminar and turbulent shear flows. Ph.D. Thesis, University Erlangen-Nürnberg (2003)
15. Özyilmaz, N.: Turbulence statistics in the inner layer of two-dimensional channel flow. M.Sci. Thesis, University Erlangen-Nürnberg (2003)
16. Lammers, P.: Direkte numerische Simulation wandgebundener Strömungen kleiner Reynoldszahlen mit dem Lattice-Boltzmann-Verfahren. Ph.D. Thesis, University Erlangen-Nürnberg (2005)
17. Lekakis, I.: HWA measurements of developed turbulent pipe flow at $Re = 50000$. private communication (2002)
18. Lammers, P., Beronov, K.N., Brenner, G., Durst, F.: Direct simulation with the lattice Boltzmann code BEST of developed turbulence in channel flows. In: Wagner, S., Hanke, W., Bode, A., Durst F. (ed) High Performance Computing in Science and Engineering, Munich 2002. Springer, Berlin (2003)
19. Beronov, K.N., Durst, F.: On the difficulties in resolving the viscous sublayer in wall-bounded turbulence. In: Friedrich, R., Geutrs, B., Métais, O. (ed) Direct and Large-Eddy Simulation V. Springer, Berlin (2004)
20. Jiménez, J., del Álamo, J.C.: Computing turbulent channels at experimental Reynolds numbers. In Proc. 15. Austral. Fluid Mech. Conf. (`www.aeromech.usyd.edu.au/15afmc/proceedings/`), Sydney (2004)

Direct Numerical Simulation of Shear Flow Phenomena on Parallel Vector Computers

Andreas Babucke, Jens Linn, Markus Kloker, and Ulrich Rist

Institute of Aerodynamics and Gasdynamics, University of Stuttgart,
Pfaffenwaldring 21, D-70569 Stuttgart, Germany,
babucke@iag.uni-stuttgart.de

Abstract A new code for direct numerical simulations solving the complete compressible 3-D Navier-Stokes equations is presented. The scheme is based on 6th-order compact finite differences and a spectral ansatz in spanwise direction. A hybrid MPI/shared-memory parallelization is implemented to utilize modern parallel vector computers as provided by HLRS. Domain decomposition and modular boundary conditions allow the application to various problems while keeping a high vectorization for favourable computing performance. The flow chosen for first computations is a mixing layer which may serve as a model flow for the initial part of a jet. The aim of the project is to learn more on the mechanisms of sound generation.

1 Introduction

The parallel vector computers NEC SX-6 and NEC SX-8 recently installed at HLRS led to the development of a new code for spatial direct numerical simulations (DNS) of the unsteady compressible three-dimensional Navier-Stokes equations. DNS simulations require high order schemes in space and time to resolve all relevant scales while keeping an acceptable number of grid-points. The numerical scheme of the code is based on the previous compressible code at the Institut für Aero- und Gasdynamik (IAG) and has been further improved by using fully 6th-order compact finite differences in both streamwise (x) and normal (y) direction. Computing the second derivatives directly leads to a better resolution of the viscous terms. By the means of grid transformation in the x-y plane one can go beyond an equidistant cartesian grid to arbitrary two-dimensional geometries. The parallelization concept of both MPI and shared memory parallelization allows to use parallel vector machines efficiently. Combining domain decomposition and grid transformation enhances the range of applications furthermore. Different boundary conditions can be applied easily due to their modular design.

The verified code is applied to a plane subsonic mixing layer consisting of two streams with unequal velocities. The intention is to model the initial part of a high Reynolds number jet and to investigate the process of sound generation inside a mixing layer. By understanding its mechanisms, we want to influence the flow in order to reduce the emitted noise. Aeroacoustic computations face the problems of i) the large extent of the acoustic field compared to the flow field and ii) the low amplitudes of the emitted sound relative to the instability waves' pressure fluctuations inside the shear region. Therefore, a high-order accurate numerical scheme and appropriate boundary conditions have to be used to minimize spurious numerical sound.

2 Computational Scheme

2.1 Governing Equations

The DNS code is based on the Navier-Stokes equations for 3-d unsteady compressible flows. In what follows, velocities are normalized by the inflow velocity U_∞ and all other quantities by their inflow values, marked with the subscript $_\infty$. Length scales are made dimensionless with a reference length L and time t with L/U_∞. Symbols are defined as follows: x, y and z are the spatial coordinates in streamwise, normal and spanwise direction, respectively. The three velocity components in these directions are described by u, v, w. ρ, T and p representing density, temperature and pressure. The specific heats c_p and c_v are assumed to be constant and therefore their ratio $\kappa = c_p/c_v$ is constant as well. Temperature dependance of viscosity μ is modelled using the Sutherland law:

$$\mu(T) = T^{3/2} \cdot \frac{T_\infty + T_s}{T + T_s} \tag{1}$$

with $T_s = 110.4\,\text{K}$. Thermal conductivity ϑ is obtained by assuming a constant Prandtl number $Pr = c_p\mu/\vartheta$. The most characteristic parameters describing a compressible viscous flow field are the Mach number Ma and the Reynolds number $Re = \rho_\infty U_\infty L/\mu_\infty$.

We use the conservative formulation described in [13] which results in the solution vector $\mathbf{Q} = [\rho, \rho u, \rho v, \rho w, E]$ containing the density, the three momentum densities and the total energy per volume

$$E = \rho \cdot c_v \cdot T + \frac{\rho}{2} \cdot \left(u^2 + v^2 + w^2\right) . \tag{2}$$

The continuity equation, the three momentum equations and the energy equation can be written in vector notation

$$\frac{\partial \mathbf{Q}}{\partial t} + \frac{\partial \mathbf{F}}{\partial x} + \frac{\partial \mathbf{G}}{\partial y} + \frac{\partial \mathbf{H}}{\partial z} = 0 \tag{3}$$

with the flux vectors **F**, **G** and **H**:

$$\mathbf{F} = \begin{bmatrix} \rho u \\ \rho u^2 + p - \tau_{xx} \\ \rho uv - \tau_{xy} \\ \rho uw - \tau_{xz} \\ u(E+p) + q_x - u\tau_{xx} - v\tau_{xy} - w\tau_{xz} \end{bmatrix} \tag{4}$$

$$\mathbf{G} = \begin{bmatrix} \rho v \\ \rho uv - \tau_{xy} \\ \rho v^2 + p - \tau_{yy} \\ \rho vw - \tau_{yz} \\ v(E+p) + q_y - u\tau_{xy} - v\tau_{yy} - w\tau_{yz} \end{bmatrix} \tag{5}$$

$$\mathbf{H} = \begin{bmatrix} \rho w \\ \rho uw - \tau_{xz} \\ \rho vw - \tau_{yz} \\ \rho w^2 + p - \tau_{zz} \\ w(E+p) + q_z - u\tau_{xz} - v\tau_{yz} - w\tau_{zz} \end{bmatrix} \tag{6}$$

containing normal stresses

$$\tau_{xx} = \frac{\mu}{Re}\left(\frac{4}{3}\frac{\partial u}{\partial x} - \frac{2}{3}\frac{\partial v}{\partial y} - \frac{2}{3}\frac{\partial w}{\partial z}\right) \tag{7}$$

$$\tau_{yy} = \frac{\mu}{Re}\left(\frac{4}{3}\frac{\partial v}{\partial y} - \frac{2}{3}\frac{\partial u}{\partial x} - \frac{2}{3}\frac{\partial w}{\partial z}\right) \tag{8}$$

$$\tau_{zz} = \frac{\mu}{Re}\left(\frac{4}{3}\frac{\partial w}{\partial z} - \frac{2}{3}\frac{\partial u}{\partial x} - \frac{2}{3}\frac{\partial v}{\partial y}\right) , \tag{9}$$

shear stresses

$$\tau_{xy} = \frac{\mu}{Re}\left(\frac{\partial u}{\partial y} + \frac{\partial v}{\partial x}\right) \tag{10}$$

$$\tau_{xz} = \frac{\mu}{Re}\left(\frac{\partial u}{\partial z} + \frac{\partial w}{\partial x}\right) \tag{11}$$

$$\tau_{yz} = \frac{\mu}{Re}\left(\frac{\partial v}{\partial z} + \frac{\partial w}{\partial y}\right) \tag{12}$$

and the heat flux

$$q_x = -\frac{\vartheta}{(\kappa-1)RePrMa^2}\frac{\partial T}{\partial x} \tag{13}$$

$$q_y = -\frac{\vartheta}{(\kappa-1)RePrMa^2}\frac{\partial T}{\partial y} \tag{14}$$

$$q_z = -\frac{\vartheta}{(\kappa-1)RePrMa^2}\frac{\partial T}{\partial z} . \tag{15}$$

Closure of the equation system is provided by the ideal gas law:

$$p = \frac{1}{\kappa Ma^2} \cdot \rho T . \tag{16}$$

2.2 Grid Transformation

To be able to compute complex geometries, a grid transformation in the x-y plane as described by Anderson [1] is applied. This means that the physical x-y plane is mapped onto an equidistant computational ξ-η grid:

$$x = x(\xi, \eta) \ , \ y = y(\xi, \eta) \ . \tag{17}$$

The occurring x and y derivatives need to be transformed into derivations with respect to ξ and η

$$\frac{\partial}{\partial x} = \frac{1}{J} \left[\left(\frac{\partial}{\partial \xi} \right) \left(\frac{\partial y}{\partial \eta} \right) - \left(\frac{\partial}{\partial \eta} \right) \left(\frac{\partial y}{\partial \xi} \right) \right] \tag{18}$$

$$\frac{\partial}{\partial y} = \frac{1}{J} \left[\left(\frac{\partial}{\partial \eta} \right) \left(\frac{\partial x}{\partial \xi} \right) - \left(\frac{\partial}{\partial \xi} \right) \left(\frac{\partial x}{\partial \eta} \right) \right] \tag{19}$$

$$J = \begin{vmatrix} \frac{\partial x}{\partial \xi} & \frac{\partial y}{\partial \xi} \\ \frac{\partial x}{\partial \eta} & \frac{\partial y}{\partial \eta} \end{vmatrix} = \frac{\partial x}{\partial \xi} \cdot \frac{\partial y}{\partial \eta} - \frac{\partial y}{\partial \xi} \cdot \frac{\partial x}{\partial \eta} \tag{20}$$

with the metric coefficients $(\partial x/\partial \xi)$, $(\partial y/\partial \xi)$, $(\partial x/\partial \eta)$, $(\partial y/\partial \eta)$ and J being the determinant of the Jacobi matrix. To compute second derivatives resulting from viscous terms in the Navier-Stokes equations, Eqs. (18) and (19) are applied twice taking into account that the metric coefficients and by that also the determinant of the Jacobi matrix can be a function of ξ and η. It is possible to compute the metric coefficients and their derivatives analytically if a specific grid transformation is recognized – if not, they will be computed using 4th-order central finite differences.

2.3 Spatial Discretization

As we use a conservative formulation, convective terms are discretized as one term to better restrain conservation equations. Viscous terms are expanded because computing the second derivative results in double accuracy compared to applying the first derivative twice. The Navier-Stokes equations combined with grid transformation lead to enormous terms, e.g. plotting the energy equation requires more than ten pages. Due to that, code generation had to be done using computer algebra software like Maple [11].

The flow is assumed to be periodic in spanwise direction. Therefore we apply a spectral ansatz to compute the derivatives in z direction

$$f(x, y, z, t) = \sum_{k=-K}^{K} \hat{F}_k(x, y, t) \cdot e^{i(k\gamma z)} \tag{21}$$

with f being a flow variable, $\hat{F}_k$ its complex Fourier coefficient, K the number of spanwise modes and $i = \sqrt{-1}$. The basic spanwise wavenumber γ is given by the basic wavelength λ_z which is the width of the integration domain.

$$\gamma = \frac{2\pi}{\lambda_z} \tag{22}$$

Spanwise derivatives are computed by transforming the respective variable into Fourier space, multiplying its spectral components with the their wavenumber $(i{\cdot}k{\cdot}\gamma)$ (or square of their wavenumber for second derivatives) and transforming it back into physical space. Due to products in the Navier-Stokes equations, higher harmonic spectral modes are generated at each timestep. To suppress aliasing, only 2/3 of the maximum number of modes for a specific z-resolution is used [2].

If a two dimensional baseflow is used and disturbances of u, v, ρ, T, p are symmetric and disturbances of w are antisymmetric, flow variables are symmetric/antisymmetric with respect to $z = 0$. Therefore only half the points in spanwise direction are needed and Eq. (21) is transfered to

$$f(x,y,z,t) = F_{0r}(x,y,z,t) + 2 \cdot \sum_{k=1}^{K} F_{kr}(x,y,t) \cdot \cos(k\gamma z) \quad \text{for } f \in [u,v,\rho,T,p] \tag{23}$$

$$f(x,y,z,t) = -2 \cdot \sum_{k=1}^{K} F_{ki}(x,y,t) \cdot \sin(k\gamma z) \quad \text{for } f \in [w]\,. \tag{24}$$

The spatial derivatives in x- and y-direction are computed using 6th order compact finite differences. Up- and downwind biased differences are applied to the convective terms which have a non-zero imaginary part of the modified wavenumber k^*_{mod}. Its alternating usage leads to carefully designed damping and by that allows the reduction of aliasing errors while keeping the favorable dispersion characteristics of a central scheme [8]. Different schemes can be chosen with respect to the current problem. The real and imaginary parts of the modified wavenumber k^*_{mod} are shown as a function of the nondimensional wavenum-

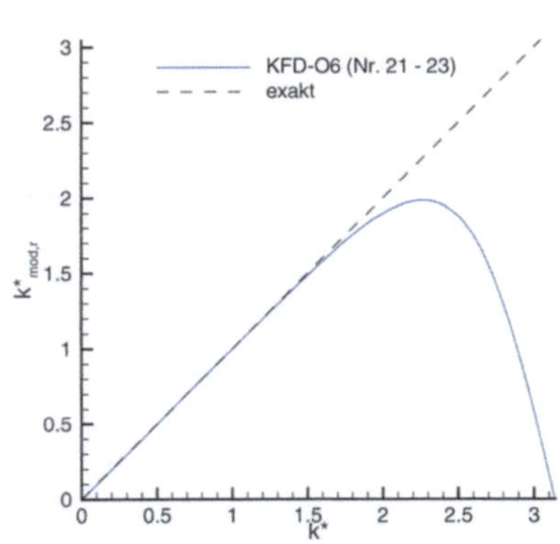

Fig. 1. Real part of the modified wavenumber $k^*_{mod,r}$, equal to k^*_{mod} of central 6th order compact finite difference

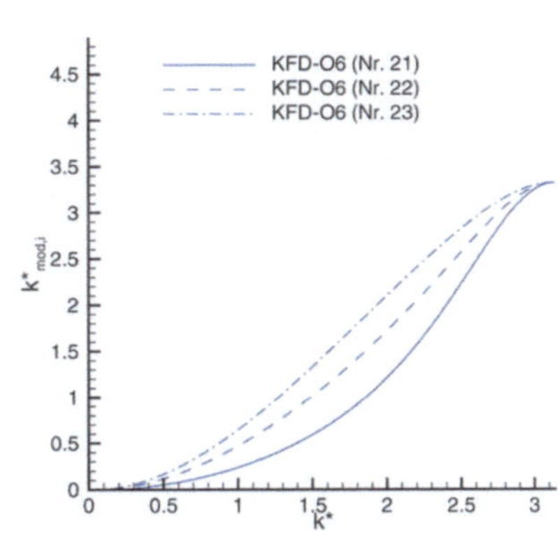

Fig. 2. Imaginary part of the modified wavenumber $k^*_{mod,i}$ for downwind biased finite differences

ber k^* in Fig. 1 and 2 for the implemented schemes. First derivatives resulting from viscous terms, caused by grid transformation and temperature dependance of viscosity, as well as second derivatives are evaluated by standard central compact FD's of 6th order. The resulting tridiagonal equation system is solved using the Thomas algorithm. The algorithm and its solution on multiple domains is discussed detailed in Sect. 2.6.

2.4 Time Integration

The time integration of the Navier-Stokes equations is done using the classical 4th order Runge-Kutta scheme as described in [8]. At each timestep and each intermediate level the biasing of the finite differences for the convective terms is changed. The ability to perform computations not only in total value but also in disturbance formulation is provided by subtracting the spatial operator of the baseflow from the time derivatives of the conservative variables $\mathbf{Q}$.

2.5 Boundary Conditions

The modular concept for boundary conditions allows the application of the code to a variety of compressible flows. Each boundary condition can either determine the primitive flow variables (u, v, w, ρ, T, p) or provide the time-derivatives of the conservative variables $\mathbf{Q}$. The spatial regime for time integration is adapted automatically. To keep the code as flexible as possible, boundary-specific parameters, such as the introduction of disturbances, are handled by the boundary conditions themselves. Up to now a variety of boundary conditions is implemented, e.g. isothermal or adiabatic walls containing a disturbance strip if specified, several outflow conditions including different damping zones or a characteristic inflow for subsonic flows having the ability to force the flow with its eigenfunctions obtained from linear stability theory.

2.6 Parallelization

To use the full potential of the new vector computer at HLRS, we have chosen a hybrid parallelization of both MPI [12] and Microtasking. As shared memory parallelization, Microtasking is used along the z direction. The second branch of the parallelization is domain decomposition using MPI. Due to the fact that the Fourier transformation requires data over the whole spanwise direction, a domain decomposition in z direction would have caused high communication costs. Therefore domain decomposition takes place in the ξ-η plane. The arbitrary domain configuration in combination with grid transformation, allows computations for a wide range of problems, e.g. the simulation of a flow over a cavity as sketched in Fig. 3.

The evaluation of the compact finite differences, described in Sect. 2.3 requires to solve a tridiagonal equation system of the form

$$a_k \cdot x_{k-1} + b_k \cdot x_k + c_k \cdot x_{k+1} = f_k \tag{25}$$

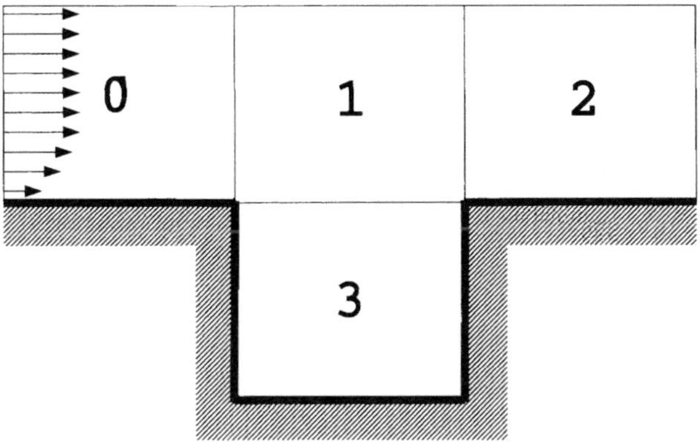

Fig. 3. Exemplary domain configuration for computation of flow over a cavity consisting of four domains. Hatched areas mark noslip wall boundary conditions

for both ξ and η direction with a, b, c being its coefficients. The computation of the RHS f is based on non-blocking `MPIISEND/MPIIRECV` communication [12]. The standard procedure for the solution of Eq. (25) is the Thomas algorithm consisting of three steps:

1. Forward-loop of LHS:

$$\begin{aligned} d_1 &= b_1 \\ d_k &= b_k - a_k \cdot \frac{c_{k-1}}{d_{k-1}} \quad , \quad (k = 2, ..., K) \end{aligned} \tag{26}$$

2. Forward-loop of RHS:

$$\begin{aligned} g_1 &= \frac{f_1}{d_1} \\ g_k &= \frac{-a_k \cdot g_{k-1} + f_k}{d_k} \quad , \quad (k = 2, ..., K) \end{aligned} \tag{27}$$

3. Backward-loop of RHS:

$$\begin{aligned} x_K &= g_K \\ x_k &= g_k - x_{k+1} \cdot \frac{c_k}{d_k} \quad , \quad (k = (K-1), ..., 1) \end{aligned} \tag{28}$$

The forward-loop of the LHS requires only the coefficients of the equation system. This has to be done only once at the initialization of the simulation. As Eqs. (27) and (28) contain the RHS f changing at every intermediate Runge-Kutta step, the computation of forward- and backward-loop of the RHS requires a special implementation to achieve acceptable computational performance. The inherent problem regarding parallel implementation is that both loops require values from the previous step, g_{k-1} for the forward-loop and x_{k+1} for the backward-loop of the RHS (note that equation (28) goes from $(K-1)$ to 1). An ad-hoc implementation would lead to large dead times because each process has to wait until the previous one has finished. To avoid that, we make use of the fact that we have to compute not only one but up to 25 spatial derivative depending on the spatial direction. The procedure is implemented as follows: the first domain starts with the forward-loop of derivative one. After its completion, the second domain continues the computation of derivate one while the first domain starts to evaluates derivative number two simultaneously. For the following steps, the algorithm continues accordingly. The resulting pipelining is shown exemplary for the forward-loop of the RHS in Fig. 4. If communication time is neglected, the theoretical speedup for forward- and backward-loop of the RHS is expressed by:

$$speedup = \frac{m \cdot n}{m + n - 1} \tag{29}$$

with n being the number of domains in a row or column respectively and m the number of equations to be solved. Theoretical speedup and efficiency of the pipelined Thomas algorithm are shown in Fig. 5 for 25 equations as a function of the number of domains. For 30 domains, efficiency of the algorithm decreases to less than 50 percent. Note that all other computations, e.g. Fourier transformation, Navier-Stokes equations and time integration, are local for each MPI process. Therefore the efficiency of the pipelined Thomas algorithm does not af-

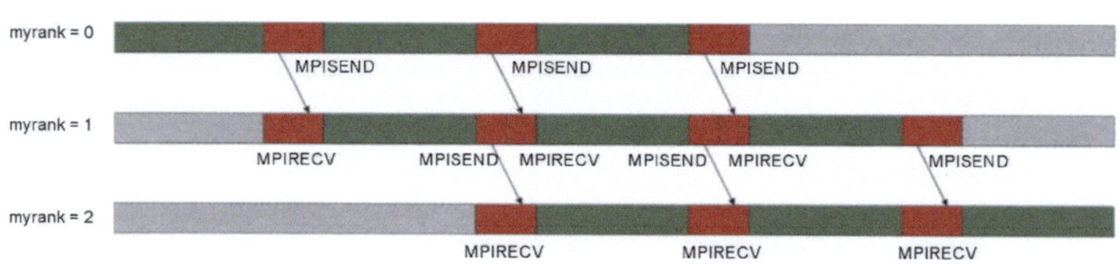

Fig. 4. Illustration of pipelining showing the forward-loop of the RHS for three spatial derivatives on three domains. Green color is denoted to computation, red to communication and grey colour shows dead time

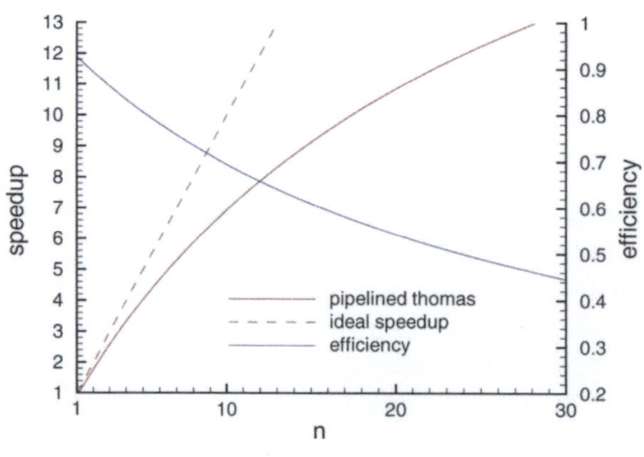

Fig. 5. Theoretical speedup and efficiency of the pipelined Thomas algorithm versus number of domains n for 25 equations

fect the speedup of the entire code that severely. The alternative to the current scheme would be an iterative solution of the equation system. The advantage would be to have no dead times, but quite a number of iterations would be necessary for a converged solution. This results in higher CPU time up to a moderate number of domains. As shared memory parallelization is implemented additionally, the number of domains corresponds to the number of nodes and therefore only a moderate number of domains will be used.

3 Verification of the Code

To verify the code, simulations of a supersonic boundary layer have been preformed and in this chapter two cases of this simulations are presented. The results from DNS are compared with *Linear Stability Theory* (LST) and with a previous results from a DNS. In the first case the linear development of a 3-d wave in a boundary layer is compared with the results from LST and in the second case results for a subharmonic resonance case are shown and compared with the work done by Thumm [13].

In both cases the Mach number is $Ma = 1.6$ and freestream temperature is $T^\star_\infty = 300\,\mathrm{K}$. A global Reynolds number of $Re = 10^5$ is chosen, which leads to a reference length scale of $L^\star = 2.963\,\mathrm{mm}$. At the lower boundary (Fig. 6) an adiabatic wall is modeled ($\frac{\partial T}{\partial y} = 0$) and at the upper boundary an exponential decay condition is used (see [13] for further details). The integration domain ends with a buffer domain, in which the disturbances are smoothly ramped to zero. Disturbances are introduced by a disturbance strip at the wall (x_{DS}) into the boundary layer. The grid resolution for both cases is the same as applied by Thumm [13]. A streamwise wave number is dissolved with 16 points, leading to a step size in $x-$direction of $\Delta x = 0.037$. The step size in $y-$direction is $\Delta y =$

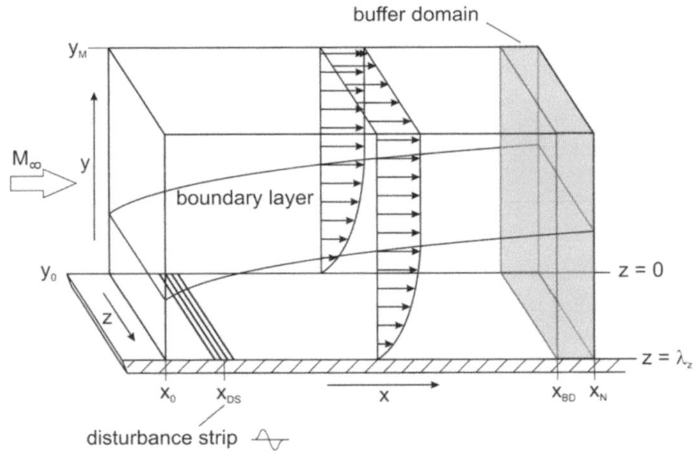

Fig. 6. Computational domain

0.00125 and two Fourier modes ($K_{max} = 2$) are employed in the z−direction. The integration domain starts at $x_0 = 0.225$ and ends at $x_N = 9.64$. The height of the domain includes approximately 2 boundary layer heights δ_{x_N} at the outflow ($y_M = 0.1$).

For a detailed investigation the flow properties are decomposed using a Fourier-decomposition with respect to the frequency and the spanwise wave number

$$\phi'(x,y,z,t) = \sum_{h=-H}^{H} \sum_{k=-K}^{K} \hat{\phi}_{(h,k)}(x,y) \cdot e^{i(h\omega_0 t + k\gamma_0 z)} \quad , \tag{30}$$

where ω_0 is the fundamental frequency at the disturbances spectrum, and $\gamma_0 = \frac{2\pi}{\lambda_z}$ the basic spanwise wave number.

3.1 Linear Stage of Transition

In this section, a 3-d wave ($\Psi = \arctan(^{\gamma}/_{\alpha_r}) \simeq 55^\circ \Rightarrow \gamma = 15.2$) with a small amplitude ($A_{(1,1)} = 5 \cdot 10^{-5}$) is generated at the disturbance strip. The development of the disturbance is linear, so the results can be compared with LST. The frequency parameter ($F = \frac{\omega}{Re} = \frac{2\pi f^* L^*}{u^*_\infty Re}$) is chosen to $F_{(1,1)} = 5.0025 \cdot 10^{-5}$. In Fig. 7 the amplification rate α_i for the u'-velocity from DNS and LST are plotted over the x-coordinate. A gap is found between the results, the amplification rates from DNS is higher then those obtained from LST. This gap is also in the simulation of Thumm [13] and Eißler [5], they shove it to non-parallel effects. Maybe this is the reason for the discrepancies in the amplification rates of the u'-velocity. In Figs. 8–10 the eigenfunctions of u', v' and p'-disturbance profiles at $x = 4.56$ from DNS and LST are shown. The agreement between DNS and LST is much better for the eigenfunctions of the 3-d wave then for the amplification rates.

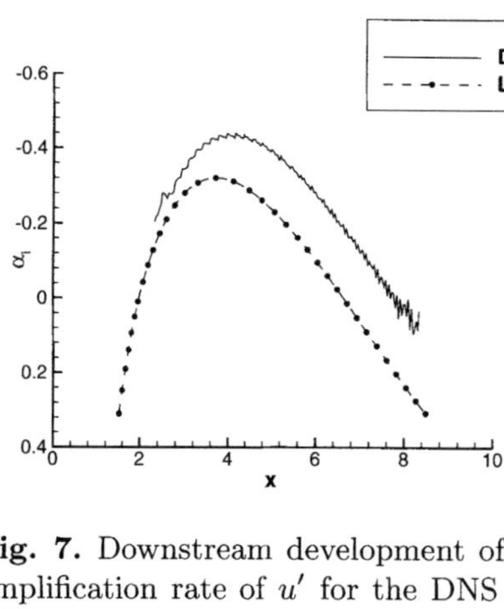

Fig. 7. Downstream development of the amplification rate of u' for the DNS and LST

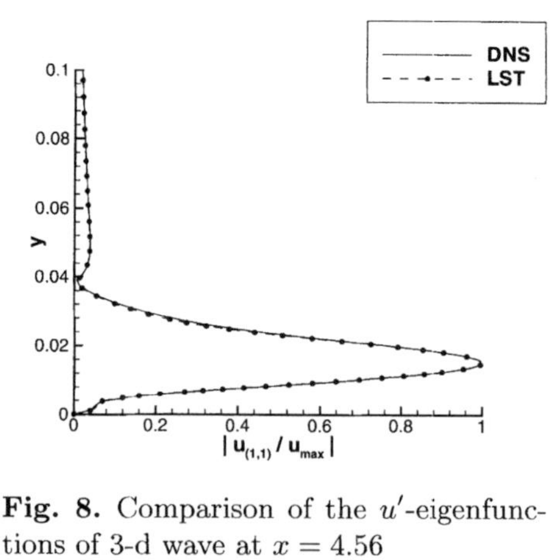

Fig. 8. Comparison of the u'-eigenfunctions of 3-d wave at $x = 4.56$

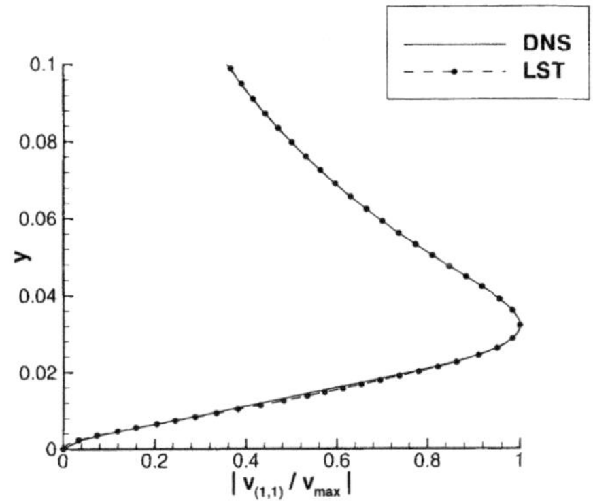

Fig. 9. Comparison of the v'-eigenfunctions of 3-d wave at $x = 4.56$

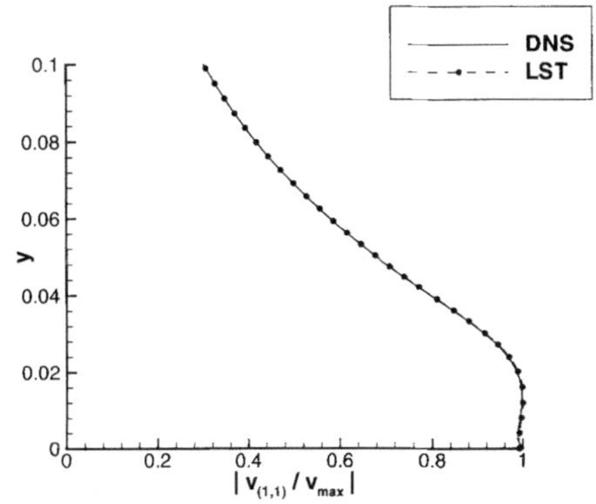

Fig. 10. Comparison of the p'-eigenfunctions of 3-d wave at $x = 4.56$

3.2 Nonlinear Stage of Transition

For the validation of the scheme in the nonlinear stage of transition a subharmonic resonance case from Thumm [13] has been simulated. The two disturbances, a 2-d and a 3-d wave ($\Psi \simeq 45° \Rightarrow \gamma = 5.3$), are now introduced into the integration domain. The frequency parameter for the 2-d wave is $F_{(1,0)} = 5.0025 \cdot 10^{-5}$ and the 3-d wave is $F_{(^1/_2,1)} = 2.5012 \cdot 10^{-5}$. The amplitudes are $A_{(1,0)} = 0.003$ and $A_{(^1/_2,1)} = 10^{-5}$. When the amplitude of the 2-d wave reaches 3–4% of the freestream velocity u_∞, the damped 3-d wave interacts non-linear with the 2-d wave and subharmonic resonance occurs (see Figs. 11–12). This means that the phase speed c_{ph} of the small disturbance adjusts to the phase speed of the high amplitude disturbance. Due to that the 3-d wave grows strongly non-linear.

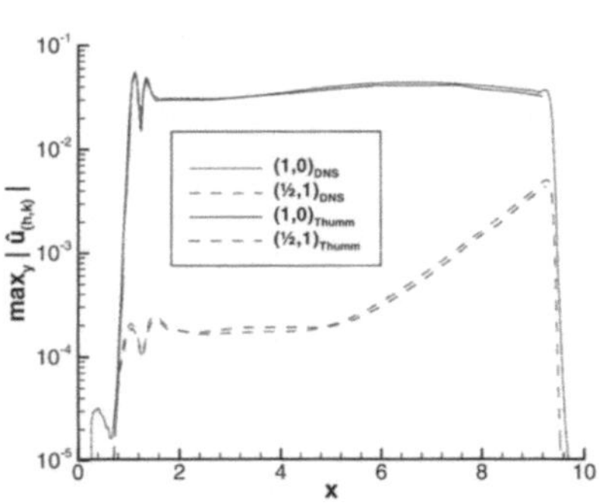

Fig. 11. Amplitude development of the u'-velocity downstream for the subharmonic resonance case

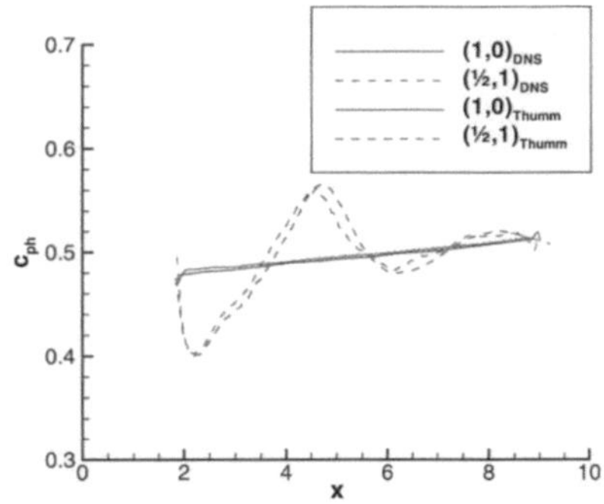

Fig. 12. Phase speed c_{ph} of the v'-velocity for the subharmonic resonance case at $y = 0.0625$

The downstream development of the u'-disturbances obtained from DNS is shown in Fig. 11, the results from Thumm [13] for this case are plotted as well. Thumms results differ only slightly. A reason for the small discrepancies is the different disturbance generation method. Thumm disturbs only v' while in the simulations here, $(\rho v)'$ is disturbed at the wall.

In Fig. 12 the phase speed of the v'-velocity over x at $y = 0.0625$ for the 2-d and 3-d wave is shown for the DNS and the results of Thumm. The phase speed of the 3-d wave approach to the 2-d wave further downstream. Although it is unknown at which y-coordinate, Thumm has determined the phase speed, the results show a good agreement.

4 Simulation of a Subsonic Mixing Layer

The current investigation is part of the DFG-CNRS project "Noise Generation in Turbulent Flows" [15]. Our motivation is to simulate both the compressible mixing layer itself as well as parts of the surrounding acoustic field. The term mixing layer describes a flow field composed of two streams with unequal velocities and serves as a model flow for the initial part of a jet as illustrated by Fig. 13. Even with increasing computational power, one is limited to jets with low Reynolds numbers [6].

4.1 Flow Parameters

The flow configuration is closely matched to the simulation of Colonius, Lele and Moin [4]. The Mach numbers are $Ma_1 = 0.5$ and $Ma_2 = 0.25$, with the subscripts $_1$ and $_2$ denoting the inflow values of the upper and the lower stream respectively. As both stream temperatures are equal ($T_1 = T_2 = 280K$), the ratio of the streamwise velocities is $U_2/U_1 = 0.5$. The Reynolds number $Re = \rho_1 U_1 \delta/\mu = 500$ is based on the flow parameters of the upper stream and the vorticity thickness δ at the inflow x_0

$$\delta(x_0) = \left(\frac{\Delta U}{|\partial u/\partial y|_{max}} \right)_{x_0} . \tag{31}$$

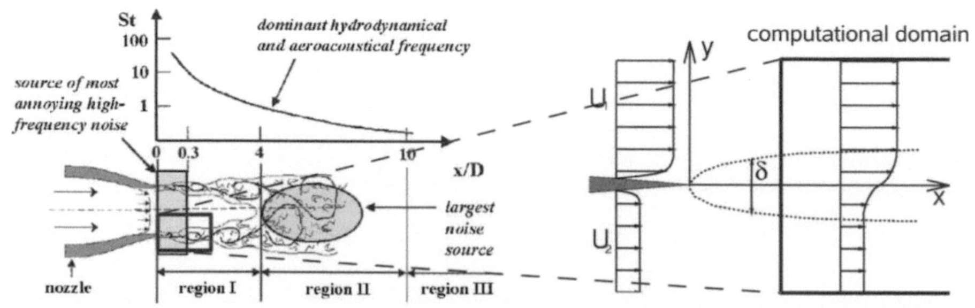

Fig. 13. Location of the computational domain showing the mixing layer as an initial part of a jet

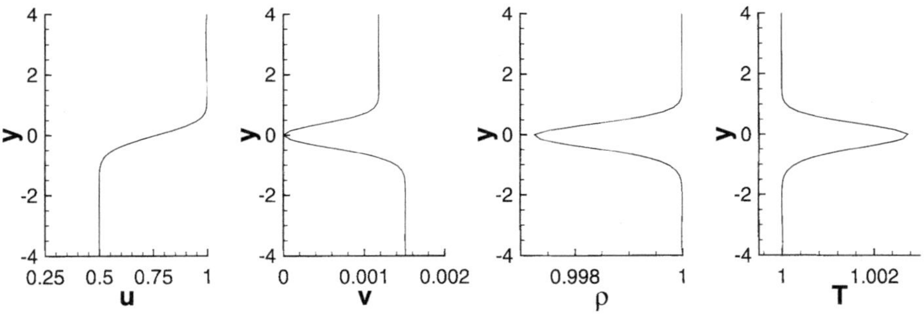

Fig. 14. Initial condition of the primitive variables u, v, ρ and T at the inflow $x_0 = 30$

The initial condition of the mixing layer is provided by solving the steady compressible two-dimensional boundary-layer equations. The initial coordinate $x_0 = 30$ is chosen in a way that the vorticity thickness at the inflow is 1. By that length scales are made dimensionless with δ. The spatial development of the vorticity thickness of the boundary layer solution is shown in Fig. 13. Velocities are normalized by $U_\infty = U_1$ and all other quantities by their values in the upper stream. Figure 14 shows the initial values at $x_0 = 30$.

A cartesian grid of 2300×850 points in x- and y-direction is used. In streamwise direction the grid is uniform with spacing $\Delta x = 0.157$ up to the sponge region where the grid is highly stretched. In normal direction the grid is continuously stretched with the smallest stepsize $\Delta y = 0.15$ inside the mixing layer ($y = 0$) and the largest spacing $\Delta y = 1.06$ at the upper and lower boundaries. In both directions smooth analytical functions are used to map the physical grid on the computational equidistant grid. The grid and its decomposition into 8 domains is illustrated in Fig. 15.

4.2 Boundary Conditions

Non-reflective boundary conditions as described by Giles [7] are implemented at the inflow and the freestream boundaries. To excite defined disturbances, the flow is forced at the inflow using eigenfunctions from linear stability theory (see Sect. 4.3) in accordance with the characteristic boundary condition. One-dimensional characteristic boundary conditions posses low reflection coefficients for low-amplitude waves as long as they impinge normal to the boundary. To minimize reflections caused by oblique acoustic waves, a damping zone is applied at the upper and lower boundary. It draws the flow variables $\mathbf{Q}$ to a steady state solution $\mathbf{Q}_0$ by modifying the time derivative obtained from the Navier-Stokes Eqs. (3):

$$\frac{\partial \mathbf{Q}}{\partial t} = \frac{\partial \mathbf{Q}}{\partial t}_{Navier\text{-}Stokes} - \sigma(y) \cdot (\mathbf{Q} - \mathbf{Q}_0) \tag{32}$$

The spatial dependance of the damping term σ allows a smooth change from no damping inside the flow field to maximum damping σ_{max} at the boundaries.

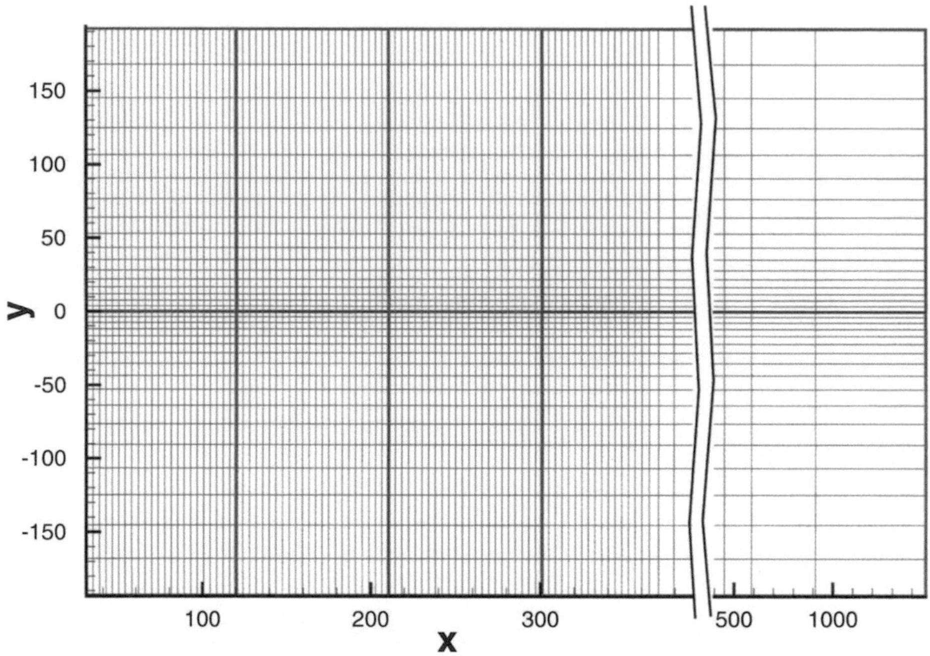

Fig. 15. Grid in physical space showing every 25th gridline. Domain decomposition in 8 subdomains is indicated by red and blue colours

To avoid large structures passing the outflow, a combination of grid stretching and low-pass filtering [9] is used as proposed by Colonius, Lele and Moin [3]. Disturbances become increasingly badly resolved as they propagate through the sponge region and by applying a spatial filter, the perturbations are substantially dissipated before they reach the outflow boundary. The filter is necessary to avoid negative group velocities which occur when the non-dimensional modified wavenumber k^*_{mod} is decreasing (see Fig. 1).

4.3 Linear Stability Theory

Viscous linear stability theory [10] describes the evolution of small amplitude disturbances in a steady baseflow. It is used for forcing of the flow at the inflow boundary. The disturbances have the form

$$\Phi = \hat{\Phi}_{(y)} \cdot e^{i(\alpha x + \gamma z - \omega t)} + c.c. \tag{33}$$

with $\Phi = (u', v', w', \rho', T', p')$ representing the set of disturbances of the primitive variables. The eigenfunctions are computed from the initial condition by combining a matrix-solver and Wielandt iteration. The stability diagram in Fig. 16 shows the amplification rates at several x positions as a function of the frequency ω. Note that negative values of $-\alpha_i$ correspond to amplification while positive values denote damping. Figure 16 shows that the highest amplification

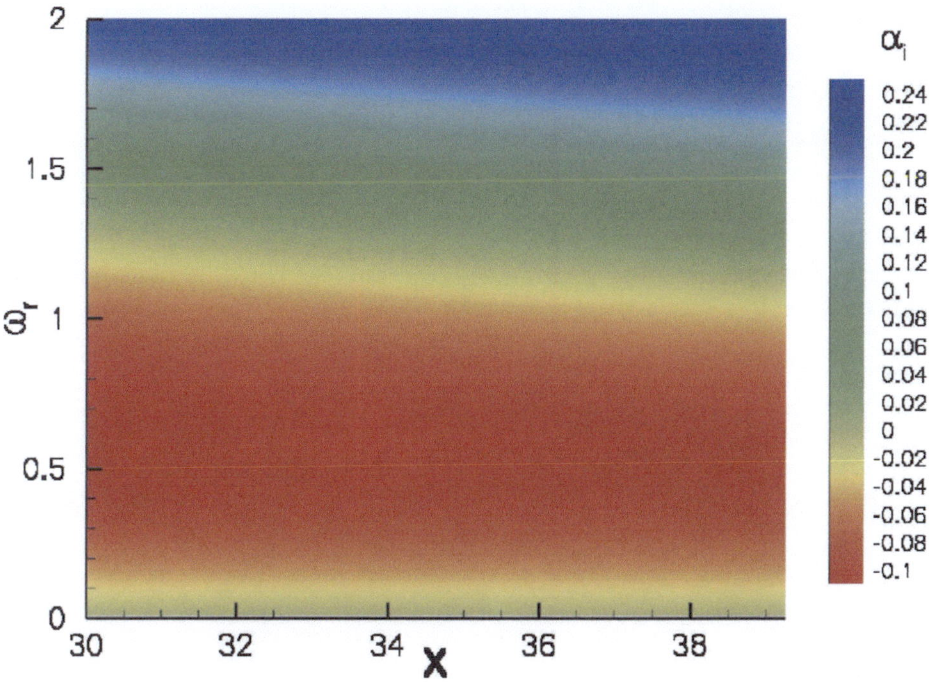

Fig. 16. Stability diagram for 2d disturbances of the mixing layer showing the amplification rate $-\alpha_i$ as a function of frequency ω and x-position

$\alpha_i = -0.1122$ is given for the fundamental frequency $\omega_0 = 0.6296$. Forcing at the inflow is done using the eigenfunctions of the fundamental frequency ω_0 and its subharmonics $\omega_0/2$, $\omega_0/4$ and $\omega_0/8$.

4.4 DNS Results

The high amplification rate as predicted by linear stability theory in the previous Sect. 4.3 leads to a soon roll-up of the mixing layer. Further downstream, vortex pairing takes place. Figure 17 illustrates the spatial development of the subsonic mixing layer by showing the spanwise vorticity. In the center of Fig. 18 ($-20 \leq y \leq 20$) the spanwise vorticity is displayed. Above and below, the dilatation $\nabla \mathbf{u}$ gives an impression of the emitted sound. At the right side, the initial part of the sponge zone is included. From the dilatation field, one can determine three major sources of sound:

- in the initial part of the mixing layer ($x = 50$)
- in the area where vortex pairing takes place ($x = 270$)
- at the beginning of the sponge region

The first source corresponds to the fundamental frequency and is the strongest source inside the flow field. Its position is upstream of the saturation of the

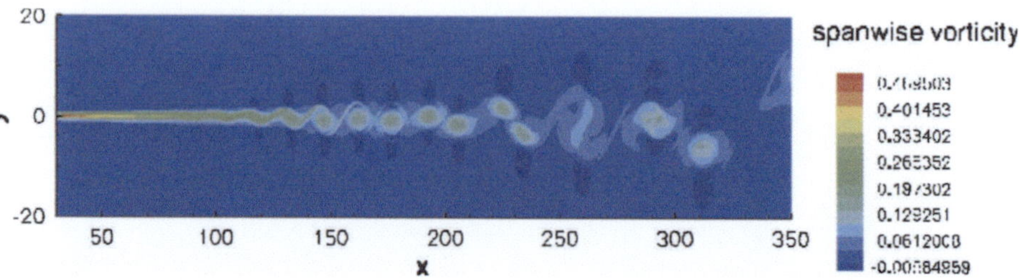

Fig. 17. Instantaneous view of the mixing layer showing roll-up of the vortices and vortex pairing by plotting spanwise vorticity

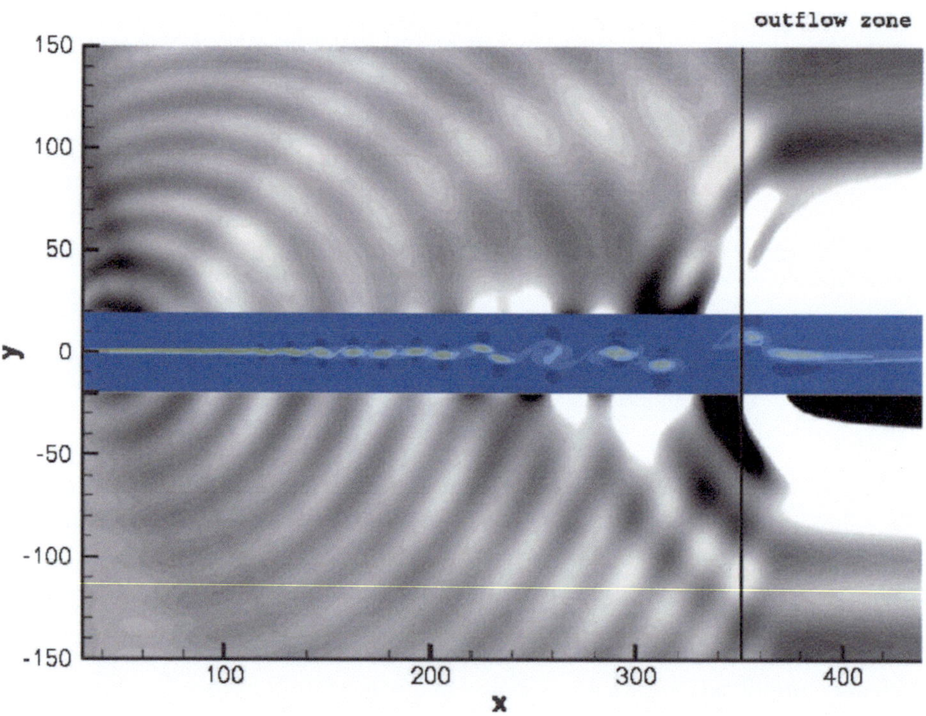

Fig. 18. Instantaneous view of the mixing layer showing spanwise vorticity in the center ($-20 \leq y \leq 20$) and dilatation to visualize the emitted sound. The beginning of the outflow zone consisting of grid-stretching and filtering is indicated by a vertical line

fundamental frequency which corresponds to the results of Colonius, Lele and Moin [4]. The second source is less intensive and therefore can only be seen by shading of the dilatation field. Source number three is directly related to the sponge zone which indicates that dissipation of the vortices occurs to fast. Due to that there is still the necessity to improve the combination of grid-stretching and filtering. As dissipation inside the outflow region is depending on the timestep Δt, choosing the appropriate combination of filter- and grid-stretching-parameters is nontrivial.

5 Performance

Good computational performance of a parallel code is first of all based on its single processor performance. As the NEC SX-8 is a vector computer we use its characteristic values for evaluation: the vector operation ratio is 99.75% and the length of the vector pipe is 240 for a 2-d computation on a grid having 575 × 425 points. Due to the fact that array sizes are already fixed at compilation, optimized memory allocation is possible which reduces the bank conflict to 2% of the total user time. All this results in a computational performance of 9548.6 MFLOP/s which corresponds to 60% of the peak performance of the NEC SX-8 [14]. Computing 30000 timesteps required a user time of 5725 seconds, so one timestep takes roughly 0.78 µs per grid-point.

To evaluate the quality of the parallelization, speedup and efficiency are taken into account. Again 30000 timesteps are computed and the grid size of each domain is those mentioned above. Figure 19 shows the dependance of speedup and efficiency on the number of MPI processes. The efficiency decreases to 83% for 8 processes. A somehow strange behaviour is the fact that the efficiency of the single processor run is less than one. Therefore efficiency is based on the maximum performance per processor. The reason for that is the non-exclusive usage of a node for runs with less than 8 processors. So computational performance can be affected by applications of other users. Comparing the achieved efficiency of 89.3% for four processors with the theoretical value of 78.1% according to Eq. (29) shows that even for 2-d computations, solving the tridiagonal equation system is not the major part of computation.

If we extend the simulation to the three-dimensional case, Microtasking, the second branch of the parallelization, is applied. We still use eight domains and by

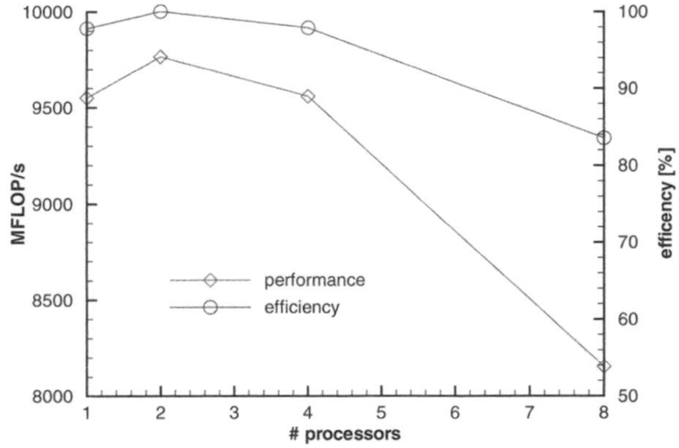

Fig. 19. Computational performance per processor (red) and efficiency (blue) as a function of MPI processes for 2-d computations

that eight MPI-processes with the same grid-size in x- and y-direction but now the spanwise direction is resolved with 33 points corresponding to 22 spanwise modes in the symmetrical case. Each MPI-process runs on its own node having 8 tasks. Computing again 30000 timesteps gives a performance of 380 GFLOP/s and by that an efficiency of 60%. One reason for the decrease in performance is the small number of spanwise modes. Best load-balancing can be achieved for a high number of spanwise modes because the z-resolution in physical space has to be of the form $2^{(kexp+1)}$ with $kexp$ depending on the number of spanwise modes. But the main reason is the poor performance of the FFT routines. Therefore we plan to implement the machine-specific MathKeisan routines. They already showed large improvements in the incompressible code N3D of IAG.

6 Outlook

A new DNS code for the unsteady three-dimensional compressible Navier-Stokes equations has been developed. An improved numerical scheme, based on the previous compressible IAG code, as well as a hybrid parallelization, consisting of MPI and shared memory parallelization, has been implemented. This allows its application to a variety of problems in compressible fluid dynamics while achieving at the same time high computational performance ($\approx$ 9 GFLOP/s per CPU). The main characteristics of the code are the following:

- solution of the full compressible three-dimensional Navier-Stokes equations
- 6th-order accurate compact finite differences in x- and y-direction
- spectral ansatz in spanwise direction (symmetric and non-symmetric)
- direct computation of the second derivatives resulting in better resolved viscous terms
- 4th-order Runge-Kutta time integration
- computation in total value or disturbance formulation
- arbitrary grid transformation in the x-y plane
- hybrid parallelization consisting of MPI and shared memory parallelization
- applicable to a wide range of problems: sub-, trans- and supersonic

To increase the performance for three-dimensional simulations, we plan to implement the FFT routines installed on the NEC SX-8 machine. As communication is not depending on spanwise resolution, we hope that performance in 3-d computations will be as good as in the 2-d case.

The code has been tested and verified for both linear and non-linear disturbances. Comparing the results with reference cases for transitional flows showed excellent agreement. The computation of a subsonic mixing layer is intended to model the initial part of a high Reynolds number jet. By choosing appropriate boundary conditions, it is possible to compute both the flow and the surrounding acoustic field. These simulations will be extended in the future to gain more details on the mechanisms of sound generation with the intention to control jet-induced noise.

Acknowledgements

The authors would like to thank the Deutsche Forschungsgemeinschaft (DFG) for its financial support within the cooperation "Noise Generation in Turbulent Flows" and the HLRS for access to computer resources and support inside the Teraflop project.

References

1. Anderson, J. D.: Computational Fluid Dynamics. McGraw-Hill, 1995
2. Canuto, C., Hussaini, M. Y., Quarteroni, A., Zang, T. A.: Spectral methods in fluid dynamics, Springer Series of Computational Physics (1988), Springer Verlag, Berlin
3. Colonius T., Lele S. K., Moin P.: Boundary Conditions for Direct Computation of Aerodynamic Sound Generation. AIAA-Journal 31, no. 9 (1993), 1574–1582
4. Colonius T., Lele S. K., Moin P.: Sound Generation in a mixing layer. J. Fluid Mech. 330 (1997), 375–409
5. Eißler, W.: Numerische Untersuchung zum laminar-turbulenten Strömungsumschlag in Überschallgrenzschichten. Dissertation, Universität Stuttgart, 1995
6. Freund J. B.: Noise Sources in a low-Reynolds-number turbulent jet at Mach 0.9. J. Fluid Mech. 438 (2001), 277–305
7. Giles M. B.: Non-reflecting boundary conditions for Euler equation calculations. AIAA-Journal 28, no. 12 (1990), 2050–2058
8. Kloker, M. J.: A robust high-resolution split type compact FD scheme for spatial direct numerical simulation of boundary-layer transition. Applied Scientific Research 59 (1998), 353–377
9. Lele, S. K.: Compact Finite Difference Schemes with Spectral-like Resolution. J. Comp. Physics 103 (1992), 16–42
10. Mack L. M.: Boundary-layer linear stability theory. AGARD-Report 709 (1984), 3.1–3.81
11. Kofler, M.: Maple V Release 2. Addison-Wesley, 1994
12. MPI Forum Mpi A message-passing interface standard. Technical Report CS-94-230, University of Tennessee, Knoxville, 1994
13. Thumm, A.: Numerische Untersuchung zum laminar-turbulenten Grenzschichtumschlag in transsonischen Grenzschichtströmungen. Dissertation, Universität Stuttgart, 1991
14. http://www.hlrs.de/hw-access/platforms/sx8/
15. http://www.iag.uni-stuttgart.de/DFG-CNRS/

Printing: Krips bv, Meppel
Binding: Stürtz, Würzburg

MIX
Papier aus verantwortungsvollen Quellen
Paper from responsible sources
FSC® C105338

If you have any concerns about our products,
you can contact us on
ProductSafety@springernature.com

In case Publisher is established outside the EU,
the EU authorized representative is:
Springer Nature Customer Service Center GmbH
Europaplatz 3, 69115 Heidelberg, Germany

Printed by Libri Plureos GmbH
in Hamburg, Germany